How to Use This Guide

The emphasis of this Student Guide is to give the "big picture" of I-DEAS, showing how it fits together as an integrated package. The material here is broader but shallower than other training courses, which are designed to give more depth in specific application areas. This Guide will be useful to those who want a quick technical introduction to I-DEAS before specializing in a particular application. It will also be useful to users who already know one application in I-DEAS and want an overview of other applications so they'll be better able to work as a team. I-DEAS is such an extensive package that seldom will someone want to learn every application in equal detail.

There are two types of hands-on activities at the end of chapters. You will first be guided through specific tutorials built into the software that demonstrate the concepts. Workshop activities with less "hand-holding" let you apply what you have learned in a design context. Many users have found they prefer to do the activities first and then read the chapter to fill in the details. Whichever you prefer to do first, each part is important. Even if you can complete the workshops without help, there may be additional topics presented in the chapter or demonstrated in the tutorials.

Hands-on activities are placed at the end of chapters rather than interspersed within so that this Guide can serve both as an introduction and as a reference. To use this Guide as a reference, please notice the Index, and the sections "Parting Suggestions" and "Where To Go For More Information" at the end of each chapter. "Parting Suggestions" give operational hints and discuss common user errors. Many of the suggestions in this section will make more sense after you have begun to use the software. The "Where To Go For More Information" section at the end of each chapter will refer you to additional information in the online Help Library. There is also some additional reference information in the appendix.

Symbols Used in This Guide

 Throughout this Guide, the "update" symbol in the left margin is used to highlight differences or new features in the latest version of I-DEAS.

 This "idea" symbol is used to indicate tips and good ideas, many of which have come from user feedback.

Workshop Format Conventions

The start of each tutorial and workshop is indicated by the workstation picture above. Several other symbols are used as road signs to guide you through the exercises.

 A checklist of things you must do before running a workshop is indicated by the check symbol. Some workshops will use files created in previous workshops that must be present in your directory before you proceed.

 The "graduation cap" precedes lists of skills you are expected to learn in the particular tutorial or workshop.

 The construction symbol is used to indicate workshop commands and actions that you are expected to perform. The symbol to the left of the icon below will help you locate the icon within the three sections of the icon panel. (The operation of icons will be discussed in Chapter 1.)

Lines

Many commands will prompt you to enter values or to pick something on the screen as shown below.

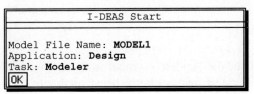

Your input is indicated in bold type. The symbols, and indicate which mouse button you should use– left, middle, or right. Descriptions of operations you are to perform such as picking graphics are given in italics.

If input in a form is requested, this Guide will present a box representing the form, showing you only what to change, not showing the exact form.

```
           I-DEAS Start

Model File Name: MODEL1
Application: Design
Task: Modeler
OK
```

 The hand shown to the left points out other places where there is an important instruction not to miss.

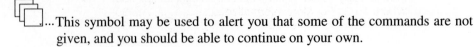 ...This symbol may be used to alert you that some of the commands are not given, and you should be able to continue on your own.

Workshop Prerequisites

Some workshops require parts from earlier workshops. These prerequisite sequences will be noted in each workshop.

Chapter 1
Introduction to I-DEAS®

I-DEAS Applications

I-DEAS® is an integrated package of mechanical engineering software tools. This software was designed to facilitate a collaborative concurrent engineering approach to mechanical engineering product design and analysis. It allows different groups in a company to share design geometry and exchange information freely for a variety of applications, many of which will be introduced in this Student Guide.

I-DEAS is composed of a number of software modules called "applications," each subdivided further into "tasks," all executed from a common user interface and sharing a common database.

You will select an application when you start I-DEAS, or from within I-DEAS you can change applications using the pull-down menus.

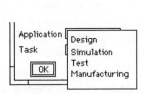

Design Application

The Design application of I-DEAS includes tasks for part modeling, drafting, assembly modeling, mechanism design, and others. There are many different ways these tools may be used together as a design evolves. For example, you may start with the outer envelope of a system of parts and work from the outside in, or you may build the parts and assemble them into a system, working from the inside out.

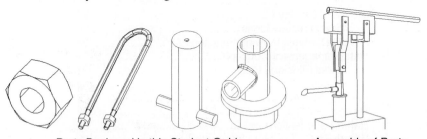

Parts Designed in this Student Guide   Assembly of Parts

Contents

Preface

The field of Mechanical Computer Aided Engineering (MCAE) is changing rapidly with advances in computer hardware and software, communication, and internet collaboration. New technology is changing the way the engineering community uses and learns MCAE tools.

What do these changes mean to how we work? In most industries, there is tremendous pressure to continually shorten the "time to market," while increasing quality at the same time. There is a major trend to do more design and analysis early in the design cycle, combined with quick prototypes to verify the design. Since design engineers spend most of their time modifying existing designs and making iterative changes, rather than starting from scratch, it is important that computer models capture not just geometry, but the design intent.

The process of design is also changing how we work together. Working in teams across multiple disciplines and physical distance requires sharing of computer-generated information. New data management tools to manage computer-transmitted information enable people to collaborate on remotely located teams. Software is also becoming more interoperable so that models can be developed and shared between multiple software systems.

Lastly, technology is changing how we learn. Although we learn in different ways, we are increasingly more impatient at learning, and don't like to read a lot of words. We expect software to be intuitive to learn and use, with visual feedback at every step so we know what the software is doing. User interface standards also help, since we can expect certain types of behavior based on using other software. Although software is becoming more sophisticated in terms of capabilities, we expect to be able to learn it in less time. Instructional media must be readily available, and be graphical and concise.

All of the above changes are reflected in the latest version of I-DEAS. This Student Guide provides a condensed overview of this family of MCAE tools. This latest version of the Student Guide is also designed to reflect these changes. The chapters are organized around the way engineers work, rather than how the software is organized. More emphasis is placed on guiding you to learn by using online tutorials and the Help Library built into the software. The workshops are designed to teach the design process, not just the function of the software.

Modeler Task

Most of the applications in the software contain the Modeler task for creating part model geometry, a flexible feature-based modeling system. The master solid model is the starting point containing the geometric definition of the parts and assemblies in a concurrent engineering design project. The solid part model can be used for many "downstream" uses such as interference studies, mass properties calculation, kinematic analysis, stress analysis, dynamics, manufacturing, testing, drafting, or other uses. The first five chapters of this Student Guide introduce some of the techniques of part modeling using this task of I-DEAS that is common to all of the applications. Later chapters will describe some of these uses and methods of sharing data among a team.

Drafting

In the framework of concurrent engineering, 2D drafting is taking the role of a "downstream" function rather than the starting point. However, having a drafting program in conjunction with a solid modeling program allows more meaningful drawings where they are needed, as in service manuals. Drawings can be made more quickly by starting with a solid model, especially when making isometric and section views, since the 3D geometry is already fully defined. Chapter 7 presents the Drafting tasks of the Design application.

Annotation that describes the manufacturing intent of a part such as GD&T and surface finish symbols is traditionally placed on 2D drawings. In I-DEAS, this notation can be placed on the 3D part model so that this information becomes part of the complete part definition, shared by other users of the 3D part model. Chapter 7 also includes part annotation because this information is important in documenting the design of the complete product.

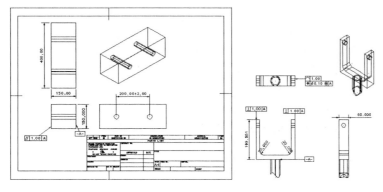

Drawings from Chapter 7

Simulation

The purpose of the Simulation application is to check for possible failure modes of parts. A variety of solution methods are available. The most common solution type is linear statics to calculate deflections and stresses, but other failure modes should be analyzed, such as fatigue and buckling. If the applied forces change rapidly, dynamic analysis should be used to calculate the response due to transient loads. If the deflections are high, non-linear analysis should be used.

The importance of simulation is not just to check that the design is good enough, but to use parameter studies and optimization to help the designers improve the design.

Since part geometry is modeled using the same Modeler task, analysts can either get parts (modeled by others) from a library, or they can model the geometry themselves using the same tools discussed in chapters 1 through 5. Although not covered in detail, Chapter 9 of this Student Guide will introduce different methods of analyzing parts using the finite element method.

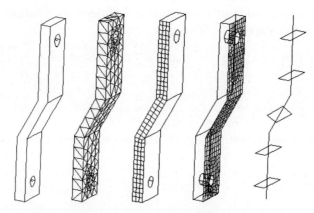

Different Types of Finite Element Models of the Same Part

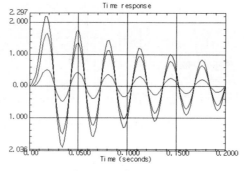

Dynamic Response

Manufacturing

The fundamental reason for designing a product is so that it can eventually be manufactured. In a traditional design cycle, manufacturing was often considered a step that came only after the design was complete. The idea of concurrent engineering is to integrate all of the different disciplines into the design phase. Using integrated tools at an early stage in the design, engineers can pose questions such as *Which process is best to produce this component? Are geometric changes required to make the product easier to manufacture?* These questions can be answered using the same part geometry which is used by the other members on the design team.

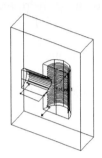

Chapter 8 of this Student Guide introduces the Manufacturing application, showing how parts can be machined using NC (Numerical Control) machining directly from the part models.

Test

In the traditional engineering design process, testing was done mainly to qualify the design after manufacturing. The goal in a concurrent engineering design process is not to eliminate testing, but to make better use of the test data by using more advanced analysis techniques. The Test application also improves communication by using graphic presentations of analysis results with an integrated tool.

The Test application offers a variety of data analysis tools, including fatigue analysis and modal testing to extract natural frequency and mode shapes. Chapter 10 will guide you to information on these tools, and other functionality of I-DEAS not covered in detail in this Student Guide.

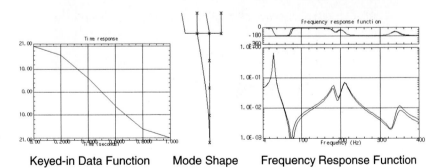

Keyed-in Data Function   Mode Shape   Frequency Response Function

Summary

In summary, the philosophy behind I-DEAS is to encourage collaborative engineering design by offering an integrated set of design automation tools. Each of the applications described in this chapter can be used by itself, but the real advantages occur when these tools are used together, allowing closer communication between different disciplines working together on a design project.

I-DEAS is only a collection of tools. The use of these tools does not guarantee good designs. It does not guarantee that the parts will not fail if the wrong loading conditions or failure conditions were analyzed. But in the hands of creative minds with the knowledge and ability to use them, these tools provide the power to take a design full-cycle, analyzing more concepts in less time, resulting in higher quality products.

The rest of this chapter will cover the specific program mechanics of using this package of MCAE design automation tools.

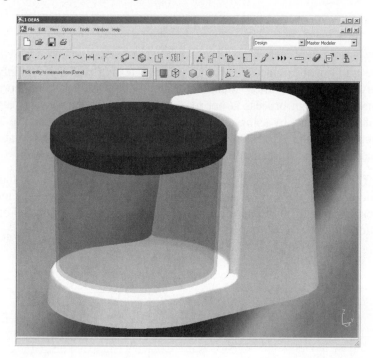

I-DEAS Software User Interface

This section covers the basic I-DEAS program mechanics. The goal is to show the consistency behind the way the software is designed, the similarity in functions across all applications, the way you interact with the software, and the way data you create is managed. Once you have learned the basics of the user interface, you will not only be more comfortable with the application you are learning, but will also be able to learn other applications at a faster pace.

Starting I-DEAS

I-DEAS uses multiple windows that you can independently move and size. You should already know how to move windows, resize them, minimize them, and display them again before you use the software.

On a Windows system, you will normally start I-DEAS using the Start, Programs menu or by clicking on an icon. On a UNIX system, type "ideas" in an operating system window unless another procedure has been set up.

On Windows, you will select either the X3D or OGL graphics display type from the Program menu when you start the software. The OGL (Open Graphics Language) display allows dynamic rotation of shaded displays if you have a graphics card that supports it. If you are using a display without hardware graphics, use "X3D," which only allows real-time rotation of wireframe graphics.

On a UNIX system, The first time you use the software, it will ask you what type of display device you are using. The software will remember what type of display device you used last time in a parameter file normally stored in your home directory. If you want to enter a different display type in a later session, you may enter the display type as a parameter on the command line as:

```
$ ideas -dOGL
```

The "-d" means that this parameter is setting the display type. There are other parameters that can be typed on this line. To find out what they are, you can use a parameter "-h", for help.

 In newer I-DEAS installations, OGL may be the only display device available.

The software presents a Start form that asks you for a project name, a model file name, and the application and task you want to use.

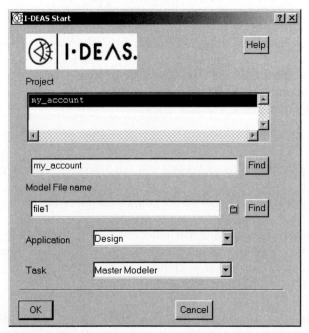

Project Name

For now, it is recommended that you leave the project name as the default, your account name. Although you may use this field to organize your work, this is not the same as using folders in the operating system. The use of projects is described in Chapter 12, Collaboration.

Model File

The model file is a database used to store your work-in-progress when you give the *Save* command. With this file, you can get back into the software later and continue where you left off. If a file of the name you enter is not found, a window will alert you that a new file will be created, so you don't mistakenly think you are opening an existing file.

✉ This name should not contain spaces or special characters. On some computers, the number of characters in the name may be limited to 10 characters. The model file actually consists of two physical files stored on the computer disk. These files have a file name you will give, plus the extensions .MF1 and .MF2. In this guide, when we refer to the model file (singular) we are really referring to both files together.

Application Menu

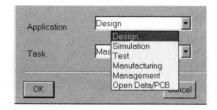

After you enter or select a project name and model file name, choose which application of I-DEAS you want. In most applications, the task you will start with is the Modeler task, as this is where geometry is entered and manipulated.

I-DEAS Windows

By default, separate window areas are used for the Graphics Window, the Prompt Window, the List Window, and the Icon Panel. To use the online tutorials in the help system, it is suggested that you resize the windows as shown to leave room for the tutorial window at the side.

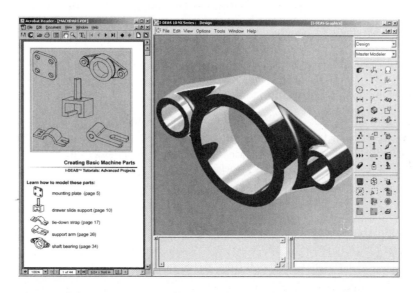

On Windows systems you may relocate toolbars as you like. Several standard toolbar layouts are available with the *Tools* menu, or you may create and store your own toolbar layouts. Although most of the exercises in this book can be done using the New User Layout as shown on page 12, the icon locators in this book and in the online tutorials refer to the Classic layout, which arranges the icons in the same positions as the icon panel on Unix systems.

Mouse Buttons

The three mouse buttons have a consistent use in each of the I-DEAS applications. The left button, sometimes called mouse button one, or just MB1, is used for most operations, such as selecting menus and icons, or picking items on the display. Most operations use one click of this button, but some operations also use a "click and drag" or a "double-click" of the left mouse button. For example, pull-down menus and icons use a click and drag operation. In a list of items on a form, sometimes a "container" name is listed, followed by ellipses (name...). To "open" this container and see its contents, use a double click.

The middle button, MB2, is the same as the Return key, often used to accept the default answer to prompts, or to end the selection of entities and perform a command.

The right button, MB3, pops up a menu of other choices to modify the mode of picking graphics. This will be further explained below in the section on Graphical Input. Hold down the right button and slide the mouse up or down to select the desired command. With the command selected, release the button. If you change your mind and don't want to make a choice, slide the mouse to the side, off the pop-up menu.

In the workshops following many of the introductory chapters in this guide, the symbols ▣□□, □▣□ and □□▣ will be used as a reminder to indicate which mouse button to use.

Mouse Button Operations

Button	Operation	Uses
▣□□	Click (Quickly)	Select icons, menus, and form entries. Pick graphic items.
	Shift Click	Pick multiple graphic items, or deselect items. Select a range of items on a form.
	Hold, Drag	Pop-up more icon choices, Pick items within a boxed area on the screen.
	Double Click	Used in forms to "open" a listed item name that is followed by ... (Not used for picking graphics.)
	Multiple Clicks (Without moving the mouse.)	"Walk" the part hierarchy. e.g.: First click picks Edge or Face, second click picks the whole Part, third click picks the Feature. The selection is indicated by highlight, a yellow bounding box on Features, or a white bounding box around the Part.
	Control Click	Turn off the Dynamic Navigator for this one pick. Select multiple items on a form.
□▣□	Click	Same as Return key. Use to pick default answer, or to end an operation.
□□▣	Hold, Drag	Select other pop-up "Cursor Menu" choices.

An acronym to help you remember the functions of the three mouse buttons is Computer-Aided Engineering (CAE) – Choose, Accept, and Extra options.

Icon Panel

Most of your command input will be by selecting icons.

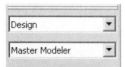

In the classic interface, the menus to switch your current application and task are found at the top of the icon panel. In the Windows OS interface, these menus are found in a floating toolbar.

The first section on the top of the icon panel contains icons that are specific to the task you are in. As you change tasks within an application, these icons will change.

The icons shown on the left are from the Modeler task. These icons are organized around the part modeling process. The first row is to select a workplane; the next three rows to sketch and dimension; the last two rows to create features.

In the Windows OS interface, these icons have a slightly different look, and are located in the *Modeler* toolbar, which may be positioned wherever you want.

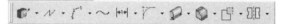

The second section contains icons that pertain to the application. These icons will generally remain fixed as you change tasks within an application. The icons shown here are the Design application icons. They include functions like *Modify*, *Move*, *Measure*, *Delete*, and *Manage*.

In the Windows OS interface, these icons are found in the *Application Commands* toolbar

The bottom section contains icons that deal with display operations that are consistent across all applications in I-DEAS. These icons control the display, such as to generate line, hidden line, and shaded image displays. They also control the view scale and viewing angle.

In the Windows OS interface, these icons are found in the *View Commands* toolbar.

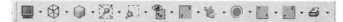

See Appendix A for a summary of the icons from each application.

Icon Operation

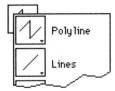

Most of the icons are actually a "stack" of related icons. To select a different operation, use the left mouse button to pull down a set of other choices. For example, to select the single line creation mode, pull down the Polyline icon as shown at left, and slide down to the (single) Lines icon.

The icon you used last will stay at the top of the "stack" of icons, so the icon panel tends to adapt to your preferences. Icons will never move to other positions in the panel, only the icon displayed on the top will change.

In the workshops in this Student Guide, to help you find icons, a small picture of the icon panel is placed next to the icon to show you where to find it in the panel. With the Windows OS interface, the icon locator will indicate which toolbar contains the icon, but not necessarily its position if you use other than the provided "Classic" toolbar layout. This picture is not part of the software; it is only an aid in this Guide. To indicate the "Move" icon in the middle section of the panel or the *Application Commands* toolbar, the workshop will show the icon as:

 Move

If this icon is not currently displayed in this position, you will need to "pull down" to select it. This operation, holding the left mouse button to "pull down" the list of icons, is very useful for a new user: the pull-down display includes the name of the icons, which will help you learn what the icons do.

Icon Summary - General Rules

The triangle in the lower right corner indicates that more icons are "below" this icon. "Pull down" with the left mouse button ▣◻◻.

The "top and side" bars over an icon indicate that this icon controls options.

A diamond shape around an icon indicates a "Manage" activity, such as to store, get, and view the directory of stored items.

The square "wraparound" shape means "modify."

Arrows to the right indicate performing an action, such as to process the changes you have described or to solve a simulation model.

Appearance. This "paint brush" icon is used to modify appearance attributes of parts and lines such as color. This "paint brush" shape appears in other icons along with other pictures, such as to modify the appearance of the workplane.

Get information, or generate a list in the list region. (Not a Help System command.)

Put something in a bin, catalog, or library. (Pointing up means "get.")

Delete.

Menu Commands

The I-DEAS user interface also has cascading menus that may optionally be used instead of, or in addition to, the Icon Panel. They can be turned on and off under the "Options" menu at the top of the icon panel, by selecting Preferences, Menus. Although there are a few advanced commands only available in the menus, most users will rarely use them. Users who want to use programmability will want to see the menu commands, as these are the foundation to programmability.

There is an equivalent menu command for most icons. In the icon reference section in Appendix A, commands are given in both icon and menu formats side by side, to serve as a reference to the menu equivalent of each icon.

Menu commands may be picked with the left mouse button or by typing at the keyboard. A menu preference option is to show the "mnemonic" at the beginning of each command. The mnemonic is particularly of interest to users who prefer to type. From the keyboard, you can also "string" commands together on one line, rather than type each line followed by a Return.

We have discussed two ways commands are given– icons and menus. Both do the same thing. In either case, the command being executed will often ask you for more information. There are three ways you will interact with commands: filling out forms, answering prompts, and graphically picking entities on the screen.

Forms Entry

Many commands will present a form to be filled out to complete the command. Forms may contain data entry fields, pull-down menus, tables, toggle switches, and radio buttons. Toggle switches used for on/off items are shown as small boxes. Clicking on these boxes turns them on ■ and off □.   Radio buttons are used where one item can be selected at a time from a list of choices. These are shown by diamond or circular buttons, one of which will be turned on ◆ (filled in) and the others turned off ◇.

Prompts

Some commands will ask for keyboard input in the Prompt Window. Prompts for keyboard input will usually start with the verb "enter..." as opposed to the verbs "pick..." implying picking graphics, or "select" to select an icon or menu command.   If a default answer to a prompt is available, it is shown in parentheses ( ) at the end of the prompt. You can either answer the question at the keyboard or accept the default answer using the Return key or the middle mouse button ⬚. In some cases, the right button ⬚ will bring up a "pop-up" menu of other choices.

To abort out of a prompt without giving an answer, select "Cancel" from the pop-up menu, or type $ at the keyboard.

Menu Commands

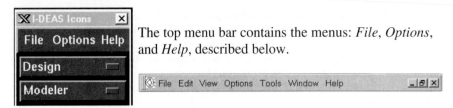

The top menu bar contains the menus: *File*, *Options*, and *Help*, described below.

File Menu

The *File* menu lets you save your work in your model file, or save to a different file using the *Save As* command. This menu also contains the important *Exit* command.

Other items under *File* are utility tools such as *Picture Files*, *Program Files*, and *Plotting*. For more information on these items, see Appendix B.

Options Menu

Two important commands in the *Options* menu are *Units* and *Preferences*. You should learn to set units and your personal preference options when you first get into a new model file.

Units

A model file has an associated system of units currently being used. All data is actually stored internally in SI units (meters and Newtons). For example, when you enter a number while you are using inch units, it will be converted to SI units internally. When this number is printed out, it will be scaled back into inch units. The active set of units should be defined when you first enter a new model file, but it can be changed at any time. For example, you can enter some quantities in inch units and others in meter units, but be careful that you are consistent. A way to by-pass units processing is to pick Meter and newton units (SI) for which all scale factors are unity.

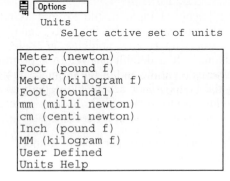

```
        Options

     Units
         Select active set of units

   Meter (newton)
   Foot (pound f)
   Meter (kilogram f)
   Foot (poundal)
   mm (milli newton)
   cm (centi newton)
   Inch (pound f)
   MM (kilogram f)
   User Defined
   Units Help

Enter system of units (mm_(milli_newton))#
```

Preferences

The *Options* menu also contains *Preferences*. These preferences include options on how icons, menus, forms, and the graphics display appear. For example, cascading menus may not be used by most users, but if you are interested in them, their display is controlled here. The workshop to follow will demonstrate how to turn menus on and off.

You can also control the attributes of the icon panel in this menu. One item you may want to control is the picture size of the icons when the stack of icons is displayed when "pulled down" by holding the left mouse button. As you are learning, it may help to make the icon picture size larger. This allows you to see each icon expanded without making the entire icon panel larger.

 In I-DEAS 8M3 or higher, use the preferences *Display, Windows Look and Feel* to select the Windows OS interface. Other new preferences control which axis you prefer to display "up" on the screen and the display of tool tips on icons and forms.

The preferences you set are stored in a file, so that the software will start each time customized to your personal choices.

Help Library

The most important skill to learn is how to use the *Help* system. The *Help* menu varies slightly between different versions of I-DEAS.

As you move the mouse over icons or forms, a brief one-line text description is given on the bottom of the graphic window. This feature is turned on and off with the *Help* menu *Quick On/Off*.

To look up detailed on-line information about an icon or form, select *Help, On Context*. The cursor will change shape to a "?". Click on the icon you would like information about. (For help on Windows forms, use the "?" on the form.) The Help Library then starts in an HTML browser, which will display the page of the reference manual describing the item you selected.

You may also navigate through all manuals directly from the bookshelf of the Help Library

The Bookshelf page of the Help Library contains links to all online
    User's Guides and
    alphabetic Command Descriptions.

The Bookshelf also contains links to self-paced tutorials and to web resources such as to the Global Technical Access Center and to the Customer Education Center.

In prior versions of I-DEAS, a higher level "Home" page contained links to the Bookshelf page, the tutorials, and to other Help Library areas. In I-DEAS version 10 or higher, all manuals, tutorials, and search are available from the Bookshelf page.

To use search, first select a book to search. For general topics, search in the I-DEAS User's Guide first, then search in specific application user guides.

For more information, click on the Help on Help icon.

Tutorials

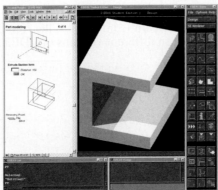

Online self-paced tutorials are available in the Help Library.

The tutorials are grouped in several different learning paths, such as "Design- Part Modeling," and "Design- Assemblies." Within each path, different levels can be selected from fundamental skills to advanced projects. Use these tutorials to supplement what you learn in this Guide.

To use the tutorials, resize the windows as shown. This will give you room to keep the tutorial right next to the I-DEAS software. Position the tutorial window in the upper left. Size its lower right corner to a vertical rectangle to fit the space left. Don't forget to press F12 or *Redisplay* after resizing the Graphics window.

Selecting Graphical Entities

As you move the mouse on the screen, you will notice entities such as points, edges, faces, and dimensions highlight when the mouse is near them. This "pre-highlighting" tells you what will happen before you click the mouse.

Commands requesting graphical input are indicated by a prompt which contains a verb like "pick" or "locate." You will then normally use the mouse to pick the entities from the graphical display window.

Pre- or Post-Selection "

A flexible feature of the I-DEAS graphical user interface is that graphical entities can be selected *before* the command is selected (pre-selection), or *after* the command is selected (post-selection). For example, you may select some lines on the screen and then press the delete icon, or select the delete icon first and then select what you want to delete. Even if no command has been selected, the cursor is active in the "picker" mode, with the arrow cursor shown above.

You may pick one or more items to process. To pick multiple items for a command, hold down the shift key while you pick. Picked items will stay highlighted.

The number of commands has been minimized in I-DEAS by the use of this flexible picking system. For example, the Move command can be used to move parts, to move the 2D workplane, or to move the text location of dimensions. In older versions, these were three different commands.

A convenient feature of pre-selection is that the items picked remain picked after the command is executed. This means if there are several parts that you want to translate and then rotate, you will not have to pick them twice if you pre-select them before you select the Move command.

If you want to clear the list of things in the "pick list," just pick where no entities are on the screen, or select the option "Deselect All" from the pop-up menu using the right mouse button ▣ .

Graphic Cursors

When the software is asking for graphical input, the cursor will change shape from an arrow to a cross-hair. Look in the Prompt Window for a description of what to pick. The cursor will change to a double cross hair when it asks you to "accept" a possible ambiguous pick. When you see the double cross-hair cursor, press the middle mouse button or return to accept the highlighted pick, or pick a different item. With any cursor displayed, you can select icon commands or "pop up" a menu of choices with the right mouse button ▣▣▪ .

▣▣▪ Reconsider (F8)

Occasionally when picking graphics, edges or faces may be behind each other, or actually coincident in space. When this happens, use the *Reconsider* option to tell the picker to cycle through a list of other possible choices. You may use this reconsider option when the cursor displays the double cross hair described above, to select among the list of items.

▣▣▪ Area Options...

Graphical entities can be picked inside or outside of an area on the screen by dragging with the left mouse button. The default method is to drag a rectangle on the screen, and all entities inside this rectangle will be picked. Other methods of picking by screen area can be chosen by selecting the menu *Area Options* from the pop-up menu, to give the form shown below.

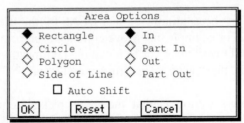

The toggle box for *Auto Shift* does the same thing as holding down the shift key to select multiple entities.

Picking and Part Hierarchy

It is important to understand the topological "hierarchy" of the parts you are picking. When picking a part on the screen, you may want to pick a point (vertex), an edge, a face, a feature, or the whole part. When you pick multiple times at the same location, the software will "walk" up different levels of the hierarchy with each pick. For example, if you pick an edge of a pocket in a part, the first pick will pick the edge or face of the part you are pointing to. The next click at the same location will pick the whole part. The next click will pick the pocket feature.

A second hierarchy in a part is the history tree that records the features applied to create the part. Features can either be picked graphically or by directly selecting the feature from the history tree. Features are accessed through the *History Access* icon or the ▣ *History Access* option. Both of these display the history tree graphically in a form which can be used to select features. The topic of part history will be more fully discussed in chapter 3.

It is important to recognize (1) the pre-highlighting before you pick an item, and (2) the graphical feedback when items are selected. Selected edges and faces will be highlighted in white. Selected features (such as holes or cut-outs) will be indicated by a dotted yellow box around the feature, and its surfaces will be highlighted in yellow. Selected parts will be indicated by a white "select box" around the part.

▣ Filter   (F11)

Another way to control what is picked is to select "*Filter...*" from the pop-up menu. This will give you a form to select the type of entities you want to make pickable.

Each command will program the filter for the types of entities it is looking for. Notice, however, that if you pick entities before selecting the command, the picker will not know what command you are going to pick next, and will allow you to pick things that are invalid for the command. As a new user, it is generally safer to select the command first, because the command will then automatically program the picker filter for the correct type of entities.

View and Display

The icons in the bottom of the icon panel control view direction and display mode (line, hidden line, or shaded displays). The view direction can also be dynamically changed using 3D Dynamic Viewing functions.

View Direction Icons

The icons in the lower left corner of the icon panel change the viewing direction to standard views. (A display preference setting changes these icons, and controls whether the Y axis or Z axis is vertical.)

3D Dynamic Viewing Functions

You can change your view direction in "real-time" with a user interface feature called "Dynamic Viewing."

The Dynamic Viewing functions are controlled with the function keys F1 through F6 and the mouse. These options are described in screen coordinates: X to the right, Y vertical, and Z toward you.

F1 controls panning the display in screen X and Y directions. Hold the function key F1 down, and move the mouse to pan the display.

F2 magnifies (zooms in or out) the scale of the display. Hold the F2 key down and move the mouse vertically on the screen.

F3 rotates the display about the three screen axes. Start with the mouse in the center of the display. While holding down F3, move the mouse vertically on the screen to rotate about the screen X axis. Move the mouse left and right to rotate about the screen Y axis. To rotate the display about the screen Z axis (perpendicular to the screen) start with the mouse near one of the corners, hold F3, and move the mouse in an arc around the center of the window.

F4 is *View Snap*, which will snap the view to the nearest orthogonal or isometric view.

F5 will reset the dynamic viewing functions back to the starting position.

Other options are available in addition to the summary provided above. If you press F6, you will be shown a function menu on the bottom of the screen. As you move the mouse pointer up and down on the screen, you will be shown other choices for the operation of functions F1-F4.

In some display modes, you may only be able to dynamically rotate line displays, but not shaded image displays. With most 3D graphic hardware, you will be able to dynamically rotate both line and shaded image displays when using a *Shaded Hardware* display.

Data Management Concepts

There are two primary areas where your parts are stored. Initially, parts are created in your model file. You may store them in libraries and catalogs either to share with others, or for your own data organization.

Sharing data with other users is covered in Chapter 12, Collaboration.

Model Files and Bins

The model file saves your working environment, including options you have set and parts you have created. Parts that are visible on the screen are described as being on your workbench. Parts in the model file can also be temporarily "put away" into "bins" in the model file, much like storing your work in a drawer in your desk.

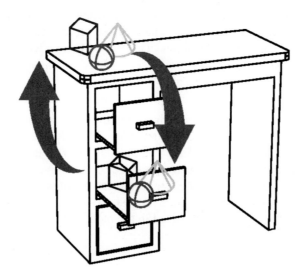

In each application of I-DEAS, there are menus to "manage" the storage and retrieval of different entities in the model file. For example, in the Modeler task, there is a *Manage Bins* icon. This command gives you a directory of the parts you have stored in the bins. It also allows you to rename, copy, and delete these parts.

A related operation is to name parts. This gives them a name and associates them with a bin, but keeps them displayed on the workbench. Naming is an important operation, since other applications can only "see" the named parts, not those that are still untitled on the workbench. When parts are named or put away, they are then tracked by the I-DEAS Data Management system. When you put a part away with the "Put Away" command, you will be prompted for a part name and bin name, if the part is not already named.

Other entities such as view options (a particular view direction, perspective, etc.), and viewport layouts (the size and locations of viewport work areas) have similar "management" concepts. There will be a Manage command with similar submenus for each of the entities stored in the database.

Although there are subtle differences between the management of different types of information, the most important thing to remember is that the Manage command determines what will be stored. The Manage Bins icon applies to parts and assemblies. The View Manage icon applies to stored view options, not parts.

 Manage Bins...   Manage Workbench Views...

While you are working, all your work is being stored in a scratch file, which contains changes from the model file. To permanently save everything you have done, use the *Save* command. This will update your model file with the changes you have made since the last save. The more often you save, the faster it will be, since the Save command is only writing the changes you have made since the last Save. If you exit and do not save, your latest work will not be saved in the model file.

Deleting Files

First, you should always use the commands in the software to delete files if possible, rather than deleting them using operating system commands. This is particularly true for model files and library data.

Items within the model file can be deleted using the *Manage Bins* command. Items in libraries can be deleted using *Manage Libraries*.

Model files are deleted in different ways, depending on what version you are using. In all versions, you may delete model files using the *Find* button on the *Start* form. (See the picture on page 14.) To delete a file, select it and pull down on the *Actions* menu to *Delete*. You can also get to this same form using the *File, Open* menu.

 In I-DEAS 8 or higher, there is a *Delete* command in the *File* menu which is the most convenient way to delete model files. There is also a new *Manage* command in the *File* menu, which allows you to perform other actions on model files, such as to rename them.

 Older versions before I-DEAS 8 have an all-purpose *Manage Items* command that can also be used to delete model files using the *Actions, Delete* menu.

Leaving I-DEAS

After saving your model file to save your work, or checking parts into a library, the command Exit will take you out of the software.

Do not try to take any short-cuts in exiting, such as exiting the window system or intentionally crashing out of the program. This may corrupt files, which could prevent you or others from getting back into the software until your System Administrator fixes the problem.

Review

This chapter covers the ground rules for running any of the I-DEAS applications. The following workshop will demonstrate some of these concepts such as how to enter and exit, how to navigate through applications and tasks, and how to select commands from icons. It is also important to learn how information is managed in the model file. You should also learn the function keys, including dynamic viewing with the function keys and the mouse.

☼ PARTING SUGGESTIONS

-A common new-user problem is holding the mouse button too long when clicking on icons. This will cause the stack of icons to momentarily "pop up," but no command will be selected.

-Learn to use the *Help* system.

-"Pull down" the icons to read the name of each, and read the Quick Help.

-Use *Option*, *Preferences*, *Icon Panel* to enlarge the pull-down picture. Use *Option*, *Preferences*, *Display* for other options.

-Don't confuse *Save* and *Put Away*.

-Don't forget to *Save* before *Exit*. Save at critical points. If you make a mistake, you can always use *File*, *Open* to get back to the last saved checkpoint. (Control-Z is a shortcut to do the same thing.)

-Do not exit by closing windows or exiting the window system. You will cause yourself more trouble than the shortcut is worth.

-Do not delete files directly from the operating system. (Use *File*, *Delete*, or *File, Open, Actions, Delete*, or *Manage Items* command, depending on your software version.)

-There are some tricks that will affect performance:

Keep your model files small by storing parts in libraries.

The program will run faster if your swap space is on a local disk. This can be changed in *Preferences*.

Although you can work with a model in shaded mode, you may get faster response for large models using line mode.

Using *Redisplay* periodically will clean up graphics and increase graphics performance.

Set the *Levels to Keep* under *Update Options* to zero or one to minimize the amount of information kept in the part history.

Where To Go For More Information

To read more about the topics described in this chapter, see:
Help, Help Library
Help on Help (icon)
I-DEAS User's Guide
Help, Help Library, Tutorials
Part Modeling – Fundamental Skills
Quick Tips to Using I-DEAS

For a reference to the major icons, see Appendix A. For troubleshooting tips, see Appendix C.

 Tutorial:

Quick Tips
to Using I-DEAS

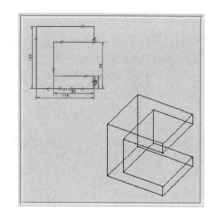

This tutorial covers some of the basic concepts from this chapter.

Before you start . . .

1. You must have an account on a computer with I-DEAS properly installed.

What you should learn:

1. The basics of the I-DEAS user interface.

2. How to use the viewing commands in the bottom section of the panel.

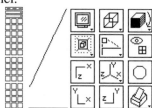

3. How to use Dynamic Viewing.

Tutorial Instructions.

Start I-DEAS with a new model file.

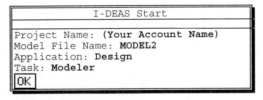

Depending on your version, select;

> *Help, Tutorials,*
> > *Quick Tips to Using I-DEAS*

> or

> *Help, Help Library, Tutorials,*
> > *Design, Design Part Modeling,*
> > > *Fundamentals,*
> > > > *Quick Tips to Using I-DEAS*

Size the graphics window and position the tutorial window in the upper left. Press the *Redisplay* icon or F12. Follow the instructions in the tutorial.

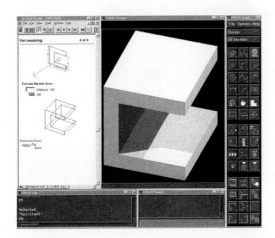

Close the tutorial window and exit from I-DEAS when you are fnished.

Workshop 1

This workshop will let you practice the concepts discussed in this chapter and demonstrated in the previous tutorial.

Before you start:

1. You must have an account on a computer or workstation that can run I-DEAS.
2. You should know how to start the software and save a model file.

After you're done, you should be able to:

1. Select I-DEAS icon commands and use the three mouse buttons.

2. Use the viewing icons in the bottom section of the panel and use the dynamic viewing controls.

3. Create basic wireframe sketches, adding and deleting wireframe geometry as needed.

4. Use the online Help Library to get help on commands.

5. Save and delete model files.

Workshop Instructions.

Start I-DEAS with a new model file.

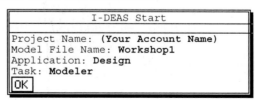

```
                    I-DEAS Start

Project Name: (Your Account Name)
Model File Name: Workshop1
Application: Design
Task: Modeler
OK
```

One of the first things you should do is define the set of units you would like to use in this model file. You do not need to do this every time, because the units you select will be stored in the model file. Pull down the *Options* menu at the top of the icon panel, and select mm units from the menu of choices. This set of units will be stored with the model file when you save.

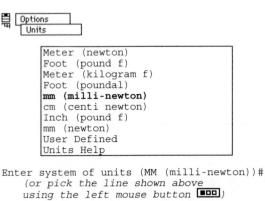

```
Enter system of units (MM (milli-newton))#   <RETURN>
      (or pick the line shown above
      using the left mouse button [■□□])
```

If you are using a Windows computer, use the Classic icon panel layout so the icon locators will match your icon positions.

```
Tools,
     Load Layout
          I-DEAS Classic Layout
```

The *Polylines* icon is at the top of the stack of related icons. Click and hold the left mouse button on the *Polylines* icon, and notice the other choices. Select the *Polylines* icon.

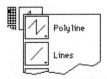

As you move the mouse, notice the digital "odometer" in the upper left corner:

```
X = 123.45
Y = 678.90
```

Instead of starting a line, press the middle mouse button, which is the same as Return, to cancel the line creation mode. The cursor will change back to an arrow, which means you can select another command or pre-select graphics to be used in another command.

```
Locate Start   [□■□]
```

Use the *Polylines* icon to sketch a shape like the following. Notice the Dynamic Navigator preview highlight changes as you move the cursor. Use the Dynamic Navigator to create vertical and horizontal lines. *Depending on how you sketch the shape, dimensions will automatically be created, which are not shown below.*

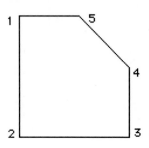

 Polylines

Locate Start	▣□□	*Pick point 1 above*
Locate End	▣□□	*Pick point 2 directly below point 1*
Locate End	▣□□	*Pick point 3 horizontal from point 2*
Locate End	▣□□	*Pick point 4 above point 3, on center of first line.*
Locate End	▣□□	*Pick point 5, aligned horizontally from point 1.*
Locate End	▣□□	*Pick point 1 again, at the start of the first line.*
Locate Start	□▣□	*(To end)*

Preselect one line and then select the *Delete* icon.

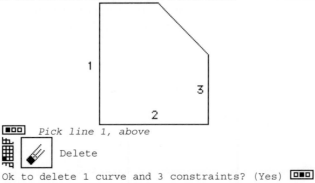

▣□□   *Pick line 1, above*

Delete

Ok to delete 1 curve and 3 constraints? (Yes) □▪□

Notice how you can pick multiple items by holding the shift key while picking.

▣□□   *Pick line 2 above*
shift-▣□□   *Hold the shift while picking line 3, above*

Delete

Ok to delete 2 curves and 1 constraint? (Yes) □▪□

This time, select the *Delete* icon first, and then pick all the remaining lines by a boxed screen area.

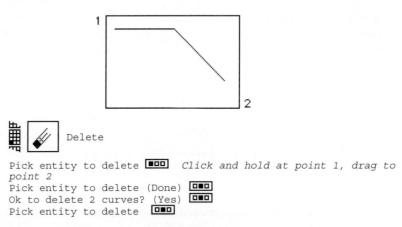

Delete

Pick entity to delete ▣□□   *Click and hold at point 1, drag to point 2*
Pick entity to delete (Done) □▪□
Ok to delete 2 curves? (Yes) □▪□
Pick entity to delete   □▪□

Most icons in I-DEAS work this way. You can "pre-select" or "post-select" items, and pick multiple items by shift-picking or by screen area. Note, however, that there is a difference in which you do first. If you select the command first, the command will set a filter to pick only the type of entities that the command is looking for. If you pick the entities first, you must make sure that you are giving the command a list of entities that it can process, or you will get an error message. It is often safer to select the command first and then pick graphic entities.

Get a block out of the Part Catalog and change the dimensions to 100 by 50 by 200 millimeters.

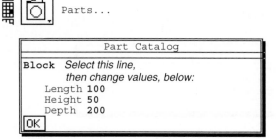

Change your viewing angle (eye position) to an isometric view. *Autoscale* the view.

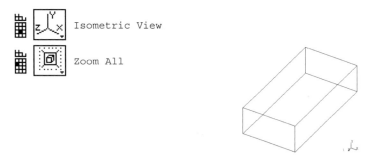

Move the cursor slowly around the vertices, edges, and faces of this block. Notice the "pre-highlighting" of the Dynamic Navigator as it highlights each of these entities.

Put this block away in the "Main" bin, giving it the name "My Block."

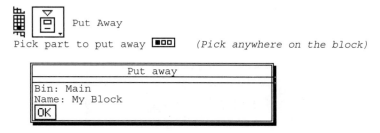

Pick part to put away ▪□□   *(Pick anywhere on the block)*

Pick part to put away □▪□

Select the *Redisplay* icon to "clean-up" the graphic display. Use this command if necessary anytime the Graphics window "doesn't look right" after deleting or moving things on the screen.

Get a cylinder from the Part Catalog, setting the dimensions to Radius=25 and Height=100.

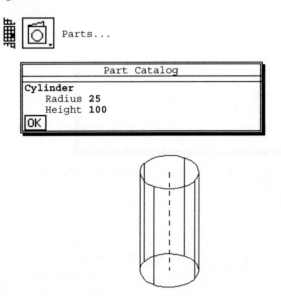

Parts...

```
                    Part Catalog
Cylinder
      Radius 25
      Height 100
OK
```

Give this part a name, but leave it on the workbench. Naming a part is similar to putting it away, in that both operations give the part a permanent name and store it in a bin. Before you do one of these operations of putting away or naming, the part is only a temporary part on the workbench. This is an important operation if you want to use the parts you create in other applications, as other applications cannot "see" the temporary parts on your workbench.

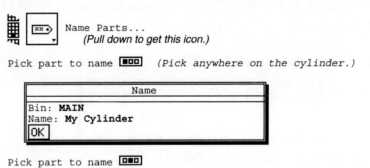

Name Parts...
    (Pull down to get this icon.)

Pick part to name ▐■□□▌ *(Pick anywhere on the cylinder.)*

```
                      Name
Bin: MAIN
Name: My Cylinder
OK
```

Pick part to name ▐□■□▌

Look at the contents of the Main bin using the *Manage Bins* icon. Notice that both parts are listed in the bin, even though one is visible on the workbench, and one has been put away and is not shown on the display. The "o" to the left of the name indicates parts that are on the workbench.

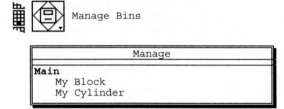

Manage Bins

Double-click on the bin name "Main." Notice that the name changes to "Main..." and the sub-items disappear from the list.

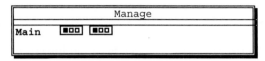

Double-click "Main..." again to open the display of the contents of the bin.

Click on the name "My Block" and then click on *Get* to get this part out on the workbench again.

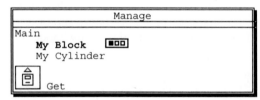

Dismiss the *Manage Bins* form.

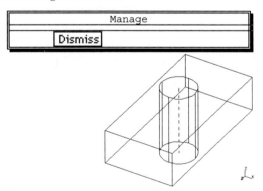

Put the cylinder away.

Put Away

Experiment with the different display types.

  Shaded

Depending on your display type, you will pick either Shaded Hardware or Shaded Software. Shaded Hardware allows dynamic rotation, shown below, if you are on a workstation with 3D graphics display capability. If you are working on an X3D display type, you will only be able to use Shaded Software, and will not be able to dynamically rotate shaded image displays.

  Hidden

  Line

Become familiar with the standard view icons in the lower left corner of the Icon Panel.

Experiment with the dynamic viewing controls described in this chapter. Depending on your display type and graphics hardware, you may be only able to perform these functions with the display in line mode.

Hold down F1 (Function 1 on the keyboard, not one of the mouse buttons), and move the mouse vertically or horizontally to pan the part on the screen.

F1   *(Hold down)*   *(Move mouse up/down, or left/right.)*

Hold down F2 and zoom the view.

F2   *(Hold down)*   *(Move mouse up/down.)*

Start with the mouse in the center of the screen and hold down F3. Move the mouse to rotate the part about the screen X and Y axes. (3D Rotation.)

F3   *(Hold down)*   *(Move mouse up/down, or left/right.)*

Start with the mouse in the corner of the screen and hold down F3. Move the mouse to rotate the part about the screen Z axes. (2D Rotation.)

F3   *(Hold down)*   *(Move mouse in arc about center.)*

Reset the view with F5.

Experiment with getting *Help*, *On Context*. The cursor will change to a question mark. Select an icon to get help on, such as *Polylines*.

On Context

The online Help Library will display the command description page, including the location of the icon, the command description, available options, and "See Also" links.

Click on one of the See Also links to see how the command is used.
(Use the Back button on the browser to get back.)

Click on the Library Bookshelf button to go to the bookshelf page of the Help Library. Open the I-DEAS User's Guide for more information on the concepts introduced in this chapter.

(If you are using a version before I-DEAS 10, use the Home button.)

Press the *Search* button, and try searching for the following words in the I-DEAS User's Guide.

Mouse
Form*
Navigator
Keyboard
Icon*

The search capability has some slight differences in different versions of I-DEAS. Read the article "Help on Help" for specific information on your version.

Look for these same topics in the index of the I-DEAS User's Guide.

If you haven't already, you should also look at the first few online tutorials. Your list of tutorials will include other tutorials not mentioned in this Student Guide that you may review to learn more.

Save your model file.

Select *Exit* from the *File* menu.

Find the model file you saved in this exercise. (Don't delete these files yet. We will show you how to delete them from within I-DEAS in the next step.)

```
model1.mf1
model1.mf2
```

Deleting Files

When you are finished using a model file, you should delete it. *(Note that the model file is referred to as singular within the software, but is actually two files on the operating system.)* This should be done from within the software, not by using operating system file delete commands. Deleting files manually from the operating system will leave the I-DEAS Data Management (IDM) system trying to track meaningless items and will lead to confusion when you try to reuse the same filenames.

Since you cannot delete the model file you are currently using, open a new file first.

 For those using I-DEAS 8 or higher, use the menu *File, Delete* to delete model files.

 For those using a version before I-DEAS 8, to delete model files use the *Find* button next to *Model File* on the Start form or on the *File, Open* form. Select any model files you don't need, and pick *Actions, Delete.*

Exit without saving this model file.

 Review:

These tutorials and this workshop covered getting in and out of the software, using icons, saving your model file, and dynamic viewing. You will need these skills in future chapters.

Chapter 2
Part Modeling

Modeler Task

The Modeler task is where the initial design concepts are developed in an integrated Mechanical Computer Aided Engineering project. The part geometry that is created here can be shared by many other "downstream" uses such as mass properties calculations, interference checking, finite element stress analysis, drafting, and manufacturing.

The geometry of a part can be created in several different ways. These include extruding or revolving a 2D section or starting from primitives (familiar solid shapes). A key philosophy in I-DEAS is to start by "sketching" sections to define the shape of the part and modify the dimensions later. I-DEAS uses a technology called "Variational Geometry" to solve the equations after dimensional or other geometric constraint changes are made.

As solid models of parts are created, features are added using construction operations which include cutting, joining, and intersecting with different parts. A die cavity can be created, for example, by cutting the part geometry from the die blank. Part geometry can be modified by changing its dimensions or by changing the features, primitives, or sections which were used to create the part. A key concept is that the software remembers the "history" of a part containing the rules that were used to create it, so that changes can be made to any operation that was performed to create the final part.

Below are some of the simple parts that we will create using the Modeler task in workshops to follow in this Student Guide.

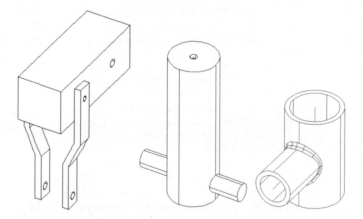

Parts Designed using Modeler Task

The Simulation and Manufacturing applications also contain the Modeler task, and have the same construction tools available to create and modify parts described in this chapter.

3D Design

New users often approach 3D feature-based solid modeling software such as I-DEAS with several misconceptions based on their past experience with other CAD software.

Drawings vs. 3D models

Users transitioning from CAD software that is primarily 2D tend to view the design process as making "drawings." A problem with this description is that "drawings" usually refer to 2D images constructed of lines. I-DEAS 3D part models use 3D lines, but the distinction is much more than just the difference between 2D and 3D lines.

I-DEAS part models are constructed as a collection of features, which may represent manufacturing process steps or functional features. In creating a "drawing," lines are drawn at the intersections of features. In the solid modeling process features are created, and the software creates the lines of intersection.

This makes a huge difference when it comes to making changes. Instead of modifying lines, the models are edited by modifying features. When features are defined and located using dimensional parameters, it is a simple matter to change these parameters.

Design vs. Modeling

Another big distinction between "drawings" and 3D solid part models is that drawings are typically used to document the final design of a product, not to actually *design* the product.

Even in I-DEAS training classes, much of the time is spent in "modeling" existing parts. New users can easily get sidetracked into thinking that modeling is design. Modeling a part from an already existing product is no more design than typing a novel from someone else's hand-written manuscript would make you an author!

As you learn I-DEAS, remember that the real work is to design a part given a set of requirements and constraints. You might not always start from scratch, because features may be reused from other designs. You will want to start a design with known datums. Position features relative to these datums using dimensional parameters that can be modified as the design evolves. Design parts in a way that maintains your design intent when parameters are changed.

Design Process

The first step of starting a part is slightly different from the design process of adding features to it. After initially creating a part, a three-step process is followed to add features.

Starting a Part

To initially create a part, consider what datums you should establish. You should also give the part a name. Make sure your units are set to the unit system you would like to use.

Although you can initially create a part without creating any datums by simply sketching a shape and extruding it, this text recommends that you use the *Create Part* icon. This command creates a starting datum coordinate system, prompts you to name the part, and prompts you to select one of the coordinate system planes to begin sketching the first feature.

   Create Part

Three-Step Process

The basic design process of adding features to parts is a repetitive three-step process:

1. Select a workplane.
2. Sketch and constrain wireframe geometry.
3. Create features.

Step 1 is to select a workplane. The workplane can be an existing face of a part, a coordinate system plane, or a reference plane.

Step 2 is called "sketching." You will sketch wireframe geometry on the selected workplane. A concept of constraints is used in the software to capture the "design intent." For example, when you sketch a line perpendicular to another line within a certain tolerance, the software will show you by its graphical feedback that it will create a perpendicular constraint. This perpendicular constraint will allow you to modify other dimensions or angles while maintaining the perpendicular relationship. Many of these constraints will be created automatically, others you may add manually.

Step 3 is to create features. Two common types of features are extrude and revolve. The software can either create a feature as a new, separate part, or combine the operations of creating the feature and applying it to an existing part.

These three steps will be repeated to add features to design a part. Different ways to perform each step will be outlined next.

Workplane

The workplane is a local coordinate system that can be moved in space. Initially, the workplane is aligned with the global coordinate system. You can change the size of the workplane display, and other attributes about it such as to display axes and a grid. The size of the workplane display is only for your visual reference, since you can sketch on the entire infinite plane.

The workplane can be manually positioned in space using the *Move* and *Rotate* commands. The workplane can also be positioned relative to other geometry using the *Align* command. Although you can directly sketch on the workplane, it is a better practice to sketch on an existing datum or feature of the part to keep features associative.

Sketch In Place

When you use the *Sketch in Place* command, the workplane will be "attached" to the face of the part, coordinate system, or reference plane that you select.

 Sketch in Place

The face you pick will be outlined in the workplane color, and the original workplane will disappear. These are graphical feedback clues to tell you that you are now working directly on the face of the part.

While aligning the workplane with the face of a part may seem similar to using the *Sketch in Place* command, there is an important difference. When you sketch in place, the geometry you sketch will belong to the part, not to the workbench. When you extrude it to create a feature, the software already knows that the feature belongs to the part, and will maintain associativity to the face used.

Note that wireframe creation commands will work in the local workplane coordinate system, wherever it is located. The only commands that use the global coordinate system are the viewing commands. For example– the icon to set the view to the front XY view is not dependent on the current position of the workplane.

Local vs. Global Coordinates

If you want to work in global coordinates, use the *Sketch on Workplane* command, then align the workplane with the global coordinate system by using *Align,* (pick the workplane), ▣▣▣ *To Global*. You will then be working on a workplane aligned with the global coordinates. If you move the workplane or *Sketch in Place*, you will then be working in local coordinates.

Rather than trying to keep track of whether you are working in a local or global mode, you can safely picture that you are always working in the workplane coordinate system, if you pay attention to the graphical feedback provided by the workplane display.

Sketching with the Dynamic Navigator™

The operation of drawing wireframe entities such as lines and circles on a workplane is called sketching. These icons are found in the second and third rows of the icon panel.

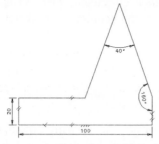

When you are sketching with any of these icons, the Dynamic Navigator™ is active to help you make logical choices and to capture your design intent. For example, when using the *Polyline* icon to create lines, while you are locating the first point for the end of the line to be created, the Dynamic Navigator will recognize things like existing end points, line centers, circle centers, other lines and curves, etc. When creating the second point, the line to be created will appear in rubberband mode, and the Dynamic Navigator will recognize perpendicular lines, parallel lines, tangencies, horizontal, vertical, as well as other points mentioned above. This saves you from having to pick from a menu at every end point creation to explicitly describe the method of point location, and minimizes the number of different line creation commands. The Dynamic Navigator also captures your design intent by creating permanent constraints where they were recognized.

It is important to learn to recognize the graphical feedback the software provides. As you create a sketch, such as shown above, the Dynamic Navigator symbols change to indicate when constraints will be created. For a table of these symbols, search for "navigator symbols" in the I-DEAS User's Guide.

Navigator Controls

If you want to change what the Dynamic Navigator can recognize, you can use the right mouse button to pop up a list of options. Selecting *Navigator...* from this menu gives you the form shown below. The first column controls what the Navigator will recognize. The second column controls whether the Navigator will also create constraints on the geometry being created to permanently enforce the conditions found. The concept of constraints will be covered in more detail in Chapter 4.

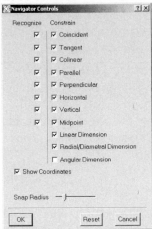

As you sketch, the symbols used to show that constraints will be created are shown in yellow. When geometric relationships are recognized, but where the software will not create a permanent constraint, the symbol will be shown in white.

Odometer

As you create geometry, you will notice a digital readout in the upper left corner of the Graphics window that gives you the cursor location, the line length, and the angle, depending on the operation. This "odometer" display can be turned off with the "Show Coordinates" toggle box in the form above.

Navigator Snap Radius

The items the Navigator can recognize in the form above will be pre-highlighted if the cursor is within a tolerance of the exact condition. This tolerance can be changed with the slider bar labeled *Snap Radius*, shown on the form, expressed as percent of screen size.

Before you start adjusting this value, if you have trouble with the Dynamic Navigator not snapping to the correct points, zoom in closer to the area of the display.

▣ Align

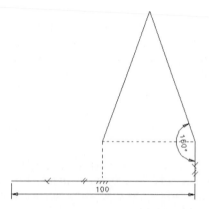

Another choice from the pop-up menu is to toggle the *Align* option. With this option on, the Dynamic Navigator will recognize endpoints and midpoints of other lines on the screen. Horizontal and vertical dashed lines will appear as you sketch when the cursor snaps to align horizontally and vertically to these locations.

▣ Focus

If you want the Dynamic Navigator to temporarily "focus" its attention on one particular curve to find relations such as perpendicular, tangent, or parallel before looking at other curves on the display, there is a *Focus* option in the pop-up menu. The software will generally try to create lines parallel and perpendicular to the first line created, as a "datum." You may control which lines you want to use for these constraints using this option.

The *Focus* option also works to project curves and points anywhere in 3D space to the plane you are sketching on. A projected point will be created as either a yellow "+" symbol or a "*" drawn the same color as the workplane (which is blue by default). The difference between the two is that the blue * will remain associative to the point focused on.

The Focus option also works when creating dimensions, to allow you to dimension to points and lines that are not on the sketch plane.

Snap Grid

Another option available when locating points on a workplane is a "snap grid," controlled by the Grid icon. This option is off by default, and is not recommended, since when you digitize on a grid, you may not capture the geometric constraints which define the design intent. Also, for most real designs, dimensions do not fall on uniformly spaced values. The grid spacing is independent from the Navigator Snap Radius, mentioned above, although both features can be on at the same time.

Sections

To create features from the wireframe geometry, I-DEAS does not require you to trim away construction lines. Instead, I-DEAS prompts you to define a section to be used in the feature. Your wireframe curves may be a nice neat closed section with each curve starting at the end of the previous curve, or you may pick curve segments out of a jumbled "haystack" of intersecting lines and curves. You may define this section in a separate command and then extrude or revolve it, or you may build the section "on the fly" when you execute the *Extrude* or *Revolve* command.

Reasons for defining a section as a separate step before the *Extrude* command might be to extract a section from another part or to modify an existing section before extruding. An example would be to extract a section from one part, offsetting it to allow for a clearance, and then use this section to cut a mating part.

There are some section options available in the pop-up menu. These include turning off the autochain function, (which is on by default) and "Stop at Intersections," which will let you pick segments of curves at each curve intersection.

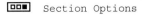   Section Options

Creating Features

A part is a collection of features. The first feature created on a part is sometimes referred to as the base feature, but there is really no difference between this first feature and other features. The process of modeling parts is the process of creating features.

Features can be created as separate parts and then applied using construction operations such as cut and join, or they can be created using the *Sketch in Place* command, which creates the feature already attached to the part in one step.

Features can be defined in different ways, such as using wireframe geometry or by getting standard parts out of a catalog. Some of the basic feature types including extrude, revolve, primitive shapes, fillet, chamfer, and shell will be described in this chapter. Other feature types requiring more than one plane of wireframe geometry or surfacing techniques will be discussed in Chapter 5.

Extrude and Revolve

The most common feature creation methods are extruding or revolving sections to get shapes like the following:

After picking the section, the *Extrude* form gives you several options to create the feature.

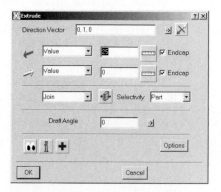

You may *Join*, *Cut*, or *Intersect* the new feature with the existing part. Intersect means to keep only the volume that lies in both the original part and the extruded shape.

The depth of the extruded feature can be defined by one of several options. Depth can be defined by a value in one direction or separate values in two directions, indicated by the green and yellow arrows. Other options are also available such as extruding until the next surface the feature intersects, or until a selected point or surface.

To flip the direction(s) of the extrusion, use the icon to invert the direction. Other options are available such as draft angle. Use help on context to read the details about the options. (Use the "?" at the top of the *Extrude* form for help on context on Windows.)

Primitives

Primitives are standard geometric shapes such as blocks, cylinders, cones, and spheres. They are an older traditional way to create features, now less used, since these shapes can all be created using extruded or revolved features. To create primitive shapes, get them out of the standard parts catalog and specify their characteristic parameters (dimensions) using the *Part...* icon to display the *Part Catalog* form.

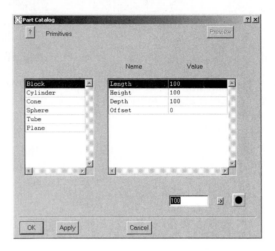

They are initially located with their center positioned at the origin of the workplane.

Below is a sample of the standard primitive shapes. The figure on the right is the same shapes with an offset added. The offset parameter is not the same as filleting the edges. Every surface is offset by this parameter.

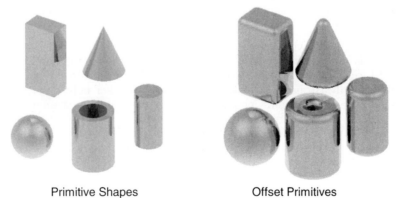

Primitive Shapes   Offset Primitives

Construction Operations

As mentioned earlier, approaching a design problem with 3D features is much different than creating drawings. For example, the extrusion on the left could be cut by an offset block to create the part in the center. Then cut with a cylinder to produce the part on the right.

There are usually many ways to make the same part, and there is no one "right" way to do it. One consideration is which method gets to the desired result with the minimum number of feature steps. This is not only faster, but will also make the part easier to modify later. Another consideration is which method most closely matches the manufacturing operation that will be used to produce the part.

The same part could be made more easily by extruding the shape including the hole as shown below. Then sketch on a face and extrude with the cut option to give the same result using only two features.

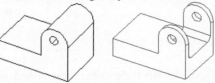

There are advantages and disadvantages to each approach. Construction operations are solid modeling tools that you should know how to use.

Cut, Join, Intersect

The commands *Cut, Join*, and *Intersect* are called "Boolean" operations. They are available as direct commands, to operate on any two existing parts. They have the same function as the extrude options cutout, protrude, and intersect.

A major choice when using these commands is whether the *Relations Switch* is turned on or off. With this switch turned off, the parts will be joined together where they exist in space. If this switch is turned on, the program will ask how to relate the two parts together before performing the Boolean operation. You have to answer more questions when performing the operation, but it will save you steps later when modifying the part, if the same positioning relationship is to be used. For example, if a cutting part was placed a fixed distance from an edge, it will maintain this fixed distance even if other dimensions are changed. If it was located as a percentage distance along an edge, it will maintain the percentage distance even if the edge length is modified.

In construction operations with relations turned on, the first prompts ask you to pick faces on two parts that will cut, join, or intersect each other. The selected faces snap together. You can then use various "face operations" to move the parts relative to each other.

If the two faces match, but one part should be flipped inside or outside the other part, use the command Flip_Faces. Another face operation is to offset one face relative to another, so that instead of just touching, one part overlaps the other.

Once the two faces are matched up, there are several positioning commands to slide the two faces relative to each other. One option is to position a point on one part along two edges from the other part using the option Along_Edges. The distance along each edge can be a fixed number or a percent of the length. Another option is From_Edges. This option lets you locate the two parts relative to one another by picking a point on one part and positioning it a fixed distance from two edges on the other part. There is also an option to position from one edge and along another. The main difference is that distances along an edge can be a percent or a fixed distance, where distances from an edge or point can only be a fixed distance. Several other options are also available, such as aligning by coincident points or setting edges parallel or at specified angles.

Orientation of Parts and Features

When parts are initially created, they are often not positioned where you want them for construction or display purposes. You can move and rotate parts using the commands *Move*, *Rotate*, and *Align*.

Parts are oriented in space by translations and rotations. The commands *Move* and *Rotate* are theoretically the only commands you need to position parts, but other commands and options may be more convenient. The *Move To* option of the *Move* command "snaps" a part so that a point picked on it will be made to coincide with a second point picked on another part. *About_Vector* rotates a part about a vector defined by picking points. *Align* matches parts face to face using similar options as those described above when using construction operations with relations turned on.

Plane Cuts

Another construction operation cuts the selected part with an infinite plane. Techniques to define a plane are described later in this chapter. The plane is drawn on the part with a "positive side" indicated by arrows. You may choose to keep the side of the part on either the positive or the negative side of the plane.

Fillets and Chamfers

The *Fillet* and *Chamfer* commands work mainly with part edges. To fillet edges, either select edges directly, or select associated vertices or surfaces to indirectly select edges to fillet. These operations are kept as part of the part history, so you will be able to go back and change the radius values later.

If you have multiple edges to fillet, remember to hold down the shift key to pick them in the same step. You can pick groups of edges with different fillet radii in the same fillet operation, resulting in one feature. There are also options for setting variable fillet radii at different locations, and changing the "conic parameter" which makes the fillet a conic rather than an arc to make smoother blends.

Fillet operations are order dependent. For example, if you fillet two edges that touch at a corner, you get a beveled intersection. If you fillet the three edges in one step, you get a "ball corner." Generally, fillets should be added as one of the last features on a part, unless there are order dependencies, such as when you want to drill a hole through a fillet.

The *Chamfer* command is very similar to *Fillet*. The difference is that the chamfer cuts straight across the edge.

Shell

The *Shell* command takes surfaces and gives them thickness. You can use this command both to start with an open surface model and create a solid, or to start with a complete solid and thicken selected surfaces, deleting others.

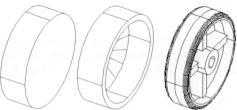

The example above shows using a shell operation to create a plastic wheel. During the shell operation, the flat face was deleted, resulting in the part in the center. After the shell operation, other features were added inside and a fillet added to the outside edges.

Reference Geometry

The technique of starting a part with a coordinate system is sometimes called the BORN (Base Orphan Reference Node) technique, because the reference coordinate, which becomes the first node in the history tree is an orphan node with no surfaces. You can create other coordinate systems, reference planes, lines, or points to use as datum geometry to locate other features. Coordinate systems are particularly useful because they provide three planes on which you can sketch features.

In the example of the wheel shown earlier, a reference plane was used to sketch the internal webs, since no face of the part existed at this location.

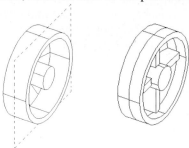

Using a Reference Plane to Sketch in Place

Defining Points, Vectors, and Planes

When creating reference geometry and also when using other commands described previously, such as *Move*, *Rotate*, and *Plane Cut*, you will need to pick points, vectors, and planes. Points are often used to define vectors, and points or vectors are used to define planes.

When the software prompts you to pick a point, remember a "pop-up" menu is available to change the mode of picking. This pop-up menu is displayed by holding down the right mouse button ▣▣▪.

The default option is to pick visible points, vectors, and planes from the display. When asked to pick a point, points at centerlines of cylinders, or intersections of edges, or the center of the workplane are pickable points. Points can also be picked off other forms of geometry such as wireframe construction geometry.

In some cases, due to previous graphical operations, you may need to draw the display again or redisplay the graphics to be able to pick. Remember these general rules: If you have trouble picking, try drawing the display again. If you still have trouble, zoom in to improve the resolution.

Also, to pick entities, they must be displayed. If points are not pickable, make sure part centerlines and centerpoints are turned on with the display filter. On some graphical displays, points may not be pickable when working in hidden line mode.

The Pop-up menu includes the following options for picking points:

Option Menu for Picking Points
  Visible
  Label
  Screen Location
  Key_in
  Intersection
  Between
  Translated
  On_curve
  On_surface
  Project_switch

If you pick the menu option *Key_In*, the prompt will change to ask you to key in the X, Y, and Z values. Selecting the option *Translated* will let you pick a point (by any of the other methods) and then key in the translation distances from the picked point to the point you would like to use. The *Between* option will let you interpolate a point between two other points. To return to the mode where you can pick a "visible" point directly, pick the option *Visible*.

A similar menu appears for defining a vector:

Option Menu for Picking Vectors
  Visible
  Key_in
  Angle
  Point_to_point
  Between
  Normal To Plane
  Heading

Picking a vector by the default *Visible* method lets you pick existing part edges or edges of the workplane to define a vector. The Heading method will let you start with a vector in the positive or negative X, Y, or Z direction, and then specify a series of angles and directions to rotate this initial vector. For example, you might start with a vector pointing in the X direction, and then rotate it 30 degrees toward the Y axis. The point-to-point method is a common method to define vectors. Note that in choosing the two points, you will have access to the entire selection of point picking options above.

To define a plane, a similar plane definition menu will be presented:

Option Menu for Picking Planes
  Visible
  Key_in
  Three_point
  Point_normal
  Offset Surface
  Axis Planes
      XY_plane
      YZ_plane
      XZ_plane

If you choose the point-normal method, you will be asked to define a point and a vector using the previous menus for point and vector definition.

Part Data Management Concepts

In Chapter 1, data management concepts were introduced using the analogy of a desk. These concepts are important to understand as you start to create multiple parts and use parts as features in constructing other parts. Parts are initially placed on the "workbench" as they are created, with a name "Untitled." More than one part can be on the workbench.

Parts can be hidden from view in the bins using the *Put Away* command, and brought back out onto the workbench using *Get*. Parts can be organized in multiple bins. By default, parts are stored in bin "Main." To create new bins to organize your stored parts, use the Create Bin button on the *Manage Bins* form.

Parts must have a name to be placed in a bin. If you use the *Put Away* command on an unnamed part, the software will ask you for a name. Naming a part enters that part as an item into the data management system. If you want to use a part in other applications, you must give it a name so that the other applications can access it. Other attributes can be added such as a part number.

There is a distinction in how a named part is treated during a construction operation. If a part used to cut or join another part does not have a name, it will be "absorbed" into the first part as a feature. If it has been given a name, a copy becomes a feature, but the original still remains as a separate part in the bin.

When parts overlap in space, it may be difficult to see how many parts are on the workbench. A useful command to use in this case is *Info*, and then pick the option *Workbench* to list the number of parts on the workbench.

Putting a part away or getting it out onto the workbench does not make copies of the part. If you want to make a copy, you must explicitly ask to make a copy of a part. If you make a mistake, you may use the *File*, *Open* command to get back to the state of the model file as of the last time you gave a *Save* command.

Summary

This chapter has covered some of the concepts that are common to the Modeler task found in most applications. The next chapter will cover specific details of modifying part geometry after it has been created.

To try some of the modeling techniques in this chapter, perform the following workshop. This workshop will illustrate basic part creation methods and construction operations described in this chapter. The parts created in this workshop will be used in several other workshops later in this Guide, so be careful about too much experimentation.

☼ PARTING_SUGGESTIONS

-Save often, to be able to correct for any mistake. Save when the software prompts you to at timed intervals unless you are following a tutorial. You can use Control-S as a shortcut to Save.

-To open your model file as it was at the last save, use the command *File, Open* (No, don't save), or just type Control-Z as a shortcut. You must terminate any command before using Control-Z.

-Don't confuse *Put Away* with *Save*. *Put Away* stores things in the scratch files, while Save makes these changes permanent in the model file.

-Parts must be given a name using *Name* or by *Put Away* before they can be accessed by other applications.

-Name parts before using them as a cutter so that they will be preserved after the cut operation.

-Deleting a part from the workbench also deletes it from the bin— use *Put Away*.

-Set your units to the dimensions you will use before creating parts.

- It's helpful to set the workplane size to get a visual reference to the size of scale.

-Pre-select parts if you need to perform multiple commands, or need to perform the same operation on multiple parts.

-To check the validity of a part, try:
  - *Info, Workbench* (Make sure you are looking at only one part.)
  - History Tree- Check the part history for extra steps.
  - List mass properties. (A closed part should have zero open surface area.)

-Use *Help, On Context* to learn more about the specific operation of commands.

Where To Go For More Information

Browse through the following online pages for more information on the topics in this chapter:

Help, Help Library   or   *Help, Help Library, Bookshelf*
     Design User's Guide
         Part Design
             Part Creation
             Part Construction

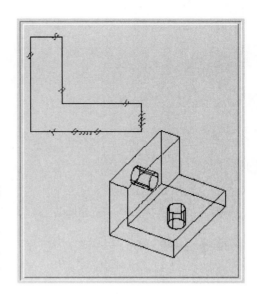

Tutorial: Creating Parts

This tutorial gives an overview of the process of creating parts. It also introduces the concept of the part history tree, which will be used more extensively in the next chapter.

Before you start:

1. You should know how to start I-DEAS and save your model file.

2. You should have a basic understanding of the user interface mechanics, including the use of the three mouse buttons and icons.

3. You should be able to use dynamic viewing to adjust your view without explicit instructions.

What you should learn:

1. The three-step process of creating part features.

2. A beginning understanding of features, and how they are recorded in the part history tree.

Tutorial Instructions.

Start I-DEAS with a new model file.

From the tutorials in the online Help Library, select:

Design, Design Part Modeling,
Fundamentals,
Creating Parts

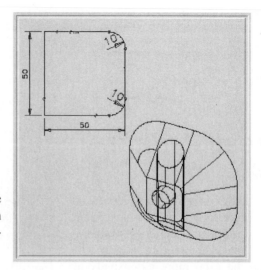

Tutorial: Extruding and Revolving Features

This tutorial goes into more depth on the most common tools used to create features- Extrude and Revolve.

Before you start:

1. You should understand the three-step modeling process.

2. You should understanding the user interface well enough to be able to read the prompts, watch the graphical feedback, and notice cursor shape changes as you work.

What you should learn:

1. The various *Extrude* options of cut and join including draft angle and corner radius options.

2. The *Revolve* options including revolve angle, translation along axis and change in radius.

Tutorial Instructions.

Select the tutorial from the online Help Library:

Design, Design Part Modeling,
Fundamentals,
Extruding and Revolving Features

After finishing the tutorial, try building the "On Your Own" parts at the end.

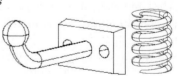

If it is available with your version of the software, continue with the first half of the tutorial "3D VGX Options on Parts." (The concepts in the second half will be covered in Chapter 4.)

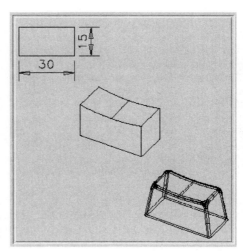

Tutorial: Adding Fillet, Shell, and Draft Features

This tutorial demonstrates an assortment of useful feature modeling techniques, modeling a computer keyboard key cap.

Before you start:

1. You should understand the three-step modeling process.

2. You should understanding the user interface well enough to be able to read the prompts, watch the graphical feedback, and notice cursor shape changes as you work.

What you should learn:

1. To use variable radius and ball-corner fillets.

2. To add draft angles directly to part faces.

3. To create thin-walled parts using the *Shell* command.

Tutorial Instructions.

Select the online Help Library tutorial:

> *Design, Design Part Modeling,*
> > *Fundamentals,*
> > > *Adding Fillet, Shell, and Draft Features*

After finishing this tutorial, you should be able to model the thin-walled plastic pencil holder in the "On Your Own" section at the end of the tutorial.

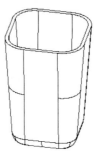

Workshop 2A: Pump

It is estimated that 40% of the six billion people living on the Earth do not have safe water to drink within a reasonable distance of their homes, and as a result, drink contaminated water. If it takes you one hour to complete this workshop, twelve hundred people will have died during that time from drinking water that is contaminated with pollutants or harmful bacteria. Many lives can be saved by the use of properly designed and installed wells. These wells provide a safe water supply as an alternative to drinking surface water contaminated by human and animal waste, agriculture, washing, or industry.

Standards of cleanliness are published by the World Health Organization on the design and installation of pumps and wells to supply safe water uncontaminated by surface run-off. A set of guidelines called Village Level Operation and Maintenance (VLOM) describes some important design considerations. The villagers using the well must (1) want it, and (2) be able to operate and maintain it on their own. The people must be educated to its importance and take ownership of it and have the ability to fix it when it breaks or needs maintenance, or they will stop using it.

The pump we will design in this workshop is patterned after a design installed in many countries by LifeWater International, a volunteer-run agency that trains people in how to drill wells in third-world countries. LifeWater's major focus is not on drilling wells, but in teaching the local people how to construct and maintain wells on their own. This pump mechanism is designed to be rugged enough to withstand rigorous use, built out of commonly available materials, and simple enough to be maintained by a rural bicycle mechanic. Rather than use ball bearings that can be contaminated by sand or water, the main pivot is made from a simple block of wood and standard pipe for bearings. It will wear out in time, but is easy to replace.

In this workshop we will develop a design concept for a hand pump mechanism as described above. The actual dimensions we will use are not a final design, but a concept to show how the parts work together. In later workshops we will show how to modify the dimensions of parts as the design is refined. A design philosophy we are illustrating is "shape then size."

For information about LifeWater International and this pump, see:

`http://www.lifewater.org`

On the main page, click on "Training," "Water Well Tutorial," then "Hand Pump."

Before you start:

1. No pre-existing files are required to run this workshop, but the model file created here will be used in later workshops.

2. It will be helpful to understand the parts of a well and pump before modeling the parts to design one. The following diagram illustrates the major parts as we will define them:

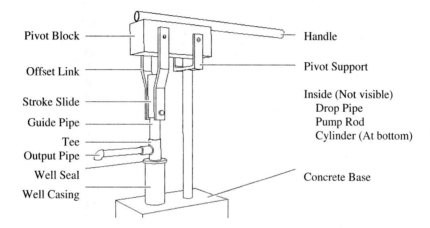

After you're done, you should be able to:

1. Create simple parts with extruded features.

2. Measure distances and move parts using these measurements.

Workshop Instructions:

Enter I-DEAS and create a new model file called "PUMP."
Enter the Design application, the Modeler task.

Create a program file if you would like to keep a record of all your commands and inputs. This program file could later be run in a new model file to re-create the parts you have made, or to demonstrate how you made them.

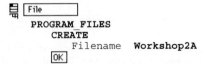
```
   File
PROGRAM FILES
    CREATE
        Filename   Workshop2A
    OK
```

From now on, until you give the command to end the creation of this file, all commands you enter will be recorded in the file Workshop2A.prg. Program files are discussed more fully in Appendix B.

Set your active units to mm.

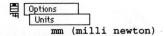
```
   Options
   Units
      mm (milli newton)
```

Define your workplane to cover an area +/- 100 mm square.

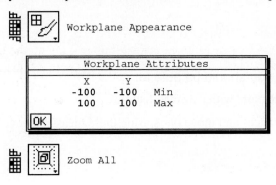 Workplane Appearance

```
          Workplane Attributes
        X        Y
      -100     -100     Min
       100      100     Max
   OK
```

       Zoom All

Create a block for pump pivot, 150 by 150, by 400 mm.

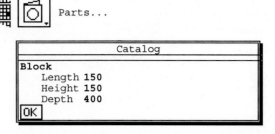

Translate this block 1000 mm up in the Y direction to put it in the approximate place it belongs in our pump assembly. View the display in an isometric view, and Zoom All to size the display area.

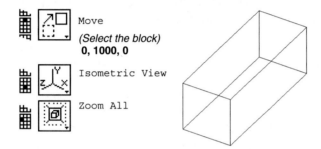

Use the Sketch in Place command to attach a workplane to the side of the block shown below. Sketch two circles aligned horizontally on the center line of the block as shown. Use the "odometer" in the upper left corner of the screen to locate them at approximately (0,-100) and (0,100) in the coordinate system of the face (where (0,0) is in the center). Use the "odometer" to make the radius approximately 15 mm. *(If you want to key in the location or radius exactly, select Options from the pop-up menu .)*

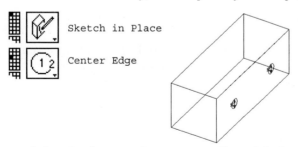

(In the workshop in the next chapter you will modify this part to set exact radius values and hole locations.)

Extrude these two circles, cutting them from the block. Pick one circle, and then add the second, to cut the two holes at the same time, This will make the two holes one "feature."

 Extrude

Pick both circles.

Cut
Thru All

If you make a mistake, what do you do?

For example, what if you joined the holes instead of cut? Or what if you forgot to select Depth, Thru All, and the holes didn't go all the way through? The solution is to pick the feature of the part, and Move it, Modify the parameters, or Delete the feature and re-do the one step.

To pick the feature, you can either (1) graphically pick it (click multiple times until the feature is highlighted with yellow corner boxes), or (2) use History Access to pick the feature from the history tree.

With the feature selected, there are several options available using the *Modify* command, such as: Feature Parameters (to re-enter the options that were used to create the feature) and Wireframe (to display the original wireframe geometry so that it can be modified).

After making modifications, remember to use the *Update* command to process the changes. *Modify* and *Update* will be discussed in Chapter 3.

 Name this part Pivot Block.

 Name Parts...

Pivot Block

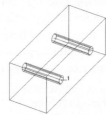

Use the *Move* command to move the Pivot Block in the Z direction so that the centerline of the hole (point 1) is over the origin, which will be on the centerline of the well.

 Move

```
Pick entity to move (Pick block)
Pick entity to move (Done) [□□■]
Enter Translation  Move To
Pick point to move from (Pick point 1, above)
Pick point to move to [□□■] Key In
Enter X,Y,Z of pt to move to (...)  ,,0
   (Change only the Z value, leaving X and Y the same)
Pick entity to move [□■□]
```

Save your work so far in your model file. (This is a good habit to develop.)

Display the block in the front view and sketch a shape as shown using the *Polyline* command to sketch the shape for the Pivot Support. This shape is being sketched on the workplane.

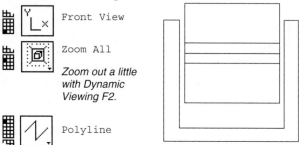

Front View

Zoom All

Zoom out a little with Dynamic Viewing F2.

Polyline

Dimensions will be included on your sketch which are not shown here, since the values are not critical as a design concept.

Use the Wireframe *Fillet* command to fillet the inside corners with a radius of 10 mm.

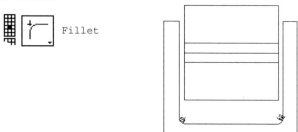

Fillet

Note: If the dimension text and arrows are sized too large, don't panic. Pre-select one dimension, then select ▣▣▣   All. Select the icon *Appearance*, and click ▣ *Autoscale*. Press *Set as Default*, and then OK.

Extrude this section 50 mm into a new part.

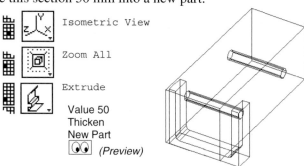

Isometric View

Zoom All

Extrude

Value 50
Thicken
New Part
👀 *(Preview)*

Sketch in place on the side of the Pivot Support. Sketch a circle with a radius of about 10 mm, closely aligned with the hole in the block. Extrude it, cutting through the part.

Sketch in Place

Center Edge

Extrude

Cut
Thru All

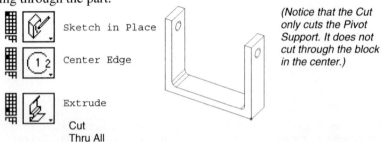

(Notice that the Cut only cuts the Pivot Support. It does not cut through the block in the center.)

Measure the distance between the hole in the Pivot Support and the right (farthest in the -Z) hole in the Pivot Block. Translate the Pivot Support by the Y and Z component of this distance to line up the holes. (You will need to zoom in using dynamic viewing to pick the circle centers.)

Measure

Distance

Move

Translation 0, DY1, DZ1

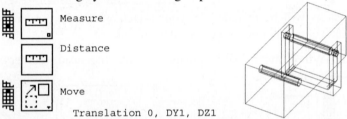

Put the Pivot Block away, leaving just the Pivot Support shown below.

Put Away

Pivot Block

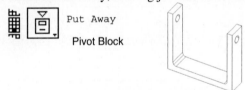

Sketch in place on the bottom of the Pivot Support. Sketch two concentric circles for the outer and inner diameter of a pipe to support it. Extrude the two circles 850 mm, into a pipe, joining to the Pivot Support. Put this part away, giving it the name "Pivot Support."

Sketch in Place

Center Edge

Extrude

Pick both circles.
Value **850**

Put Away

Pivot Support

(Use Dynamic Viewing with F1 and F2 to zoom in!)

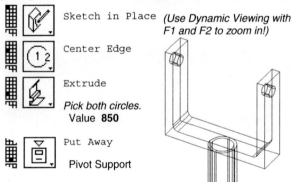

End the program file creation mode if you turned it on at the beginning of the workshop.

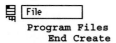

Program Files
                End Create

Save your model file.

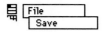

The model file is viewed as your temporary work space. A more permanent place to store parts is to put them in a project library. I-DEAS Master Series users will also use libraries to share parts between team members.

Check in the Pivot Block into a library named "Pump Parts." Check them in with the library status *Keep to modify*. The *Check In* icon on the *Manage Bins* form does the same thing as the *Check In* icon.

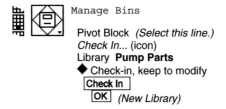

Manage Bins

Pivot Block *(Select this line.)*
Check In... (icon)
Library **Pump Parts**
◆ Check-in, keep to modify
Check In
OK *(New Library)*

Repeat the above steps to check in the Pivot Support into the same library.

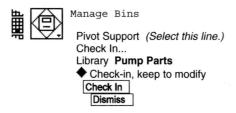

Manage Bins

Pivot Support *(Select this line.)*
Check In...
Library **Pump Parts**
◆ Check-in, keep to modify
Check In
Dismiss

This is the end of Workshop 2A. You may exit here, or continue with Workshop 2B.

Workshop 2B: Pump

This workshop will continue making parts for the pump assembly.

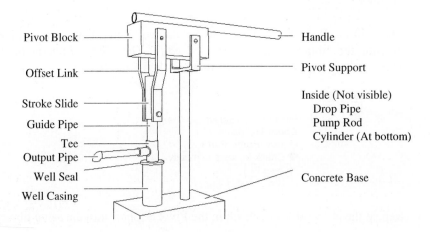

Pivot Block — Handle

Offset Link — Pivot Support

Stroke Slide
Guide Pipe
Tee
Output Pipe
Well Seal
Well Casing

Inside (Not visible)
    Drop Pipe
    Pump Rod
    Cylinder (At bottom)

Concrete Base

Before you start:

1. Start in the Design application, the Modeler task with the same model file from the last workshop.

After you're done, you should be able to

1.   Create parts with extruded and revolved features.

3.   Use a snap grid with the workplane.

Create a Program File if you want to keep a record of all your commands.

```
PROGRAM FILES
    CREATE
        Filename   Workshop2B
```

Get the Pivot block. Select the Sketch in Place command and pick the front end face of the block. Sketch a circle for the outer diameter of the pipe handle using the circle creation command Two Points On. Make the radius about 25 mm, positioned so that the bottom edge touches the top of the block.

Get

Pivot Block

Zoom All

Sketch in Place

Two Points On

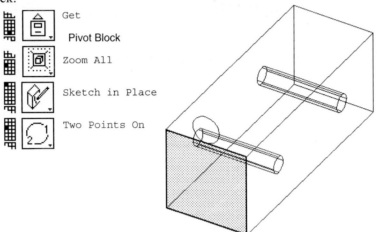

Zoom in on the area around the circle using the *Zoom* command or dynamic viewing. Use the *Center Edge* command to create a circle for the inner diameter of the pipe handle. Make the radius about 20 mm. *(Hold the control key when picking the second point, to temporarily turn off the Dynamic Navigator. If the cursor changes to a double cross-hair, and the prompt window asks you to "Accept" the pick, use the middle button to accept, then pick the second point.)*

Zoom

Center Edge

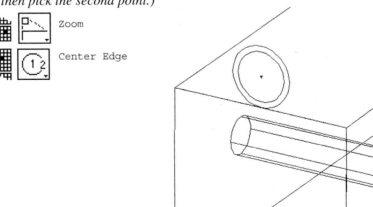

Extrude the pipe 1000 mm, making sure to select "<u>New Part</u>." Use the button "Flip Direction" to extrude in the direction shown.

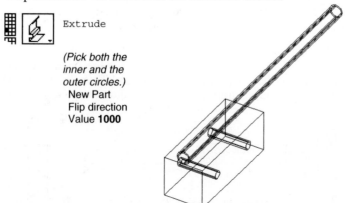

Extrude

(Pick both the inner and the outer circles.)
New Part
Flip direction
Value **1000**

Name this new part "Handle." Put away both the Handle and the Pivot Block.

Name

Handle

Put Away

Get a Tube from the Part Catalog to model the well casing with an inner radius of 45 mm, an outer radius of 50 mm, and a height of 400 mm. Translate it in the Y direction 200 mm, so that the lower end is at ground level. (We are only modeling the pump from the ground up.) Name this part "Well Casing."

Parts...

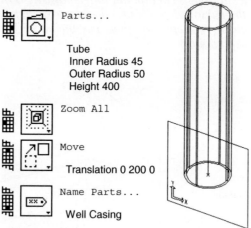

Tube
Inner Radius 45
Outer Radius 50
Height 400

Zoom All

Move

Translation 0 200 0

Name Parts...

Well Casing

Save your model file.

Save

Translate the workplane so that its origin is now located at the upper end of the well casing. Turn on the "Snap Grid" to sketch the Well Seal. (A snap grid is not often used in conceptual design, since the Dynamic Navigator is a more powerful sketching tool, but we will use it here to illustrate its use.)

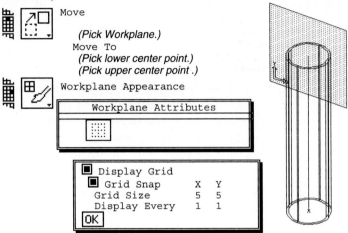

```
          Move

          (Pick Workplane.)
          Move To
          (Pick lower center point.)
          (Pick upper center point .)
          Workplane Appearance
```

```
              Workplane Attributes
```

```
    ■ Display Grid
      ■ Grid Snap      X   Y
        Grid Size       5   5
        Display Every   1   1
    OK
```

Use the *Focus* option to project the centerline of the well casing to the workplane to create a 2D wireframe line on the workplane.

```
          Polyline

          □□■ Focus
          Pick the centerline.  Zoom in if required.
          □■□
```

Zoom in and sketch the shape shown to revolve into the Well Seal. Start at point 1, go around clockwise. *(Turn off the Align option to use the grid.)*

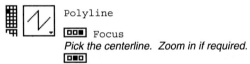

```
          Front View

          Zoom

          Polyline
          □□■ Align (off)
          Sketch points 1-7
```

Revolve this section 360 degrees about the centerline. Turn off the Grid.

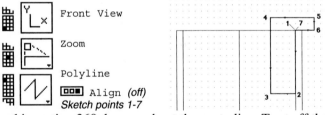

```
          Isometric View

          Revolve
          Pick section
          □■□
          Pick axis to revolve about
          □■□
          Workplane Appearance
```

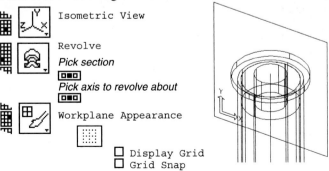

```
                    □ Display Grid
                    □ Grid Snap
```

Name this new part "Well Seal." Put away both parts.

Name Parts...
Well Seal

Put Away

If there are any wireframe curves left on the workbench, delete them.

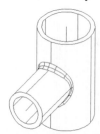

We will next model the Tee. If you think about it, you cannot create a pipe Tee section by just joining tubes. We will create a Tee from two joined cylinders to describe the outer shape, and then make extruded cuts from two directions to cut the interior shape.

Get a primitive cylinder from the parts catalog with a radius of 20 mm and a height of 80 mm. Preselect the part by clicking twice. (White boxes should box in the outer dimensions of the cylinder.) Translate this cylinder 40 mm in the Y direction, and then rotate 90 degrees about the X axis (through a point at the origin).

Parts..

Cylinder
Radius 20
Height 80

Move

Translation 0 40 0

Rotate

Pivot point (origin) ▣
About X
Angle (90) ▣

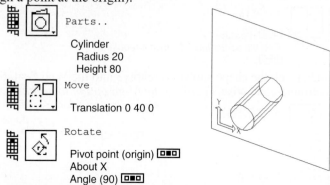

Get another primitive cylinder with a radius of 30 mm and a Height of 100 mm. Join the two together with relations turned off. (Relations will be covered in Chapter 4.)

Parts..

Cylinder
Radius 30
Height 100
▣ Deselect All

Join

▣ Turn Relations Off
Pick each cylinder

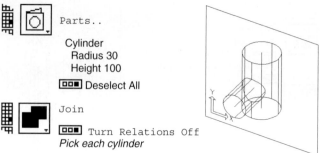

Sketch in Place on the top face of the Tee, draw a circle with a radius of about 25 mm, and extrude the circle, cutting through the Tee.

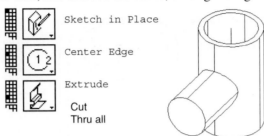

Sketch in Place

Center Edge

Extrude

Cut
Thru all

Sketch in Place on the side of the Tee, draw a circle with a 15 mm radius, and cut up to the next surface.

Sketch in Place

Center Edge

Extrude

Pick circle

Cut
Until Next

Fillet the intersection between the cylinders with a radius of 5 mm. Name this part "Tee."

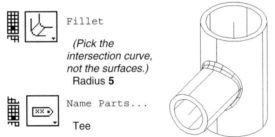

Fillet

(Pick the intersection curve, not the surfaces.)
Radius **5**

Name Parts...

Tee

Get the Well Seal. Align the Tee with the Well Seal as shown.

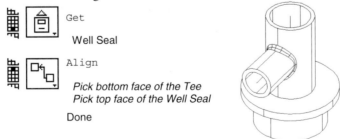

Get

Well Seal

Align

Pick bottom face of the Tee
Pick top face of the Well Seal

Done

Sketch in place on the top face of the Tee. Draw <u>one</u> circle with a radius of about 5 mm smaller than the inner radius of the hole in the Tee.

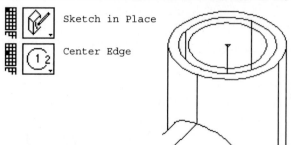

Sketch in Place

Center Edge

Extrude the existing circle of the inner diameter of the Tee and the one you just sketched into a <u>new</u> part.

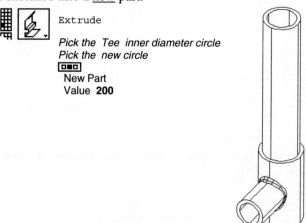

Extrude

Pick the Tee inner diameter circle
Pick the new circle

New Part
Value **200**

Name this new part "Guide Pipe." The primary purpose of this pipe is to guide the Stroke Slide we will create next so that it moves straight up and down. Put the Tee and Well Seal away.

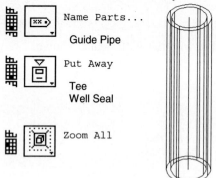

Name Parts...

Guide Pipe

Put Away

Tee
Well Seal

Zoom All

Save your work so far in your model file.

File

Save

The Stroke Slide slides over the top of the Pipe Assembly. View from the front view to sketch a section as shown. Use the *Focus* option to project the centerline of the guide pipe to the workplane to create a wireframe line on the workplane. Start in the upper left, using the "odometer" to start at a point about 5 mm off the centerline to leave a hole of 10 mm in diameter after revolving the section.

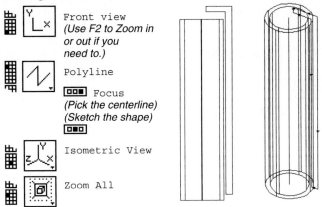

Front view
(Use F2 to Zoom in or out if you need to.)

Polyline

☐☐■ Focus
(Pick the centerline)
(Sketch the shape)
☐■☐

Isometric View

Zoom All

Revolve this section to create the Stroke Slide. Put away the Guide Pipe so that the Stroke Slide is the only part on the display.

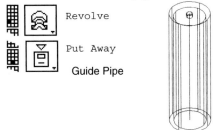

Revolve

Put Away

Guide Pipe

Sketch in place on the bottom face of the Stroke Slide to orient the workplane there. Get a primitive cylinder with a radius of 12 mm and a height of 150 mm. Because we are sketching on the face, the new part is positioned on the face, rather than at the global origin.

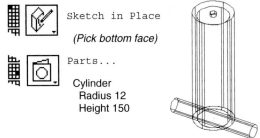

Sketch in Place

(Pick bottom face)

Parts...

Cylinder
Radius 12
Height 150

Save your model file.

File

Save

Use the *Sketch on Workplane* command so that you can translate this cylinder in global rather than in local face coordinates. Translate the cylinder 25 mm in the Y direction. Join this cylinder to the Stroke Slide. (If the prompt asks you to pick a planar face from the movable part, turn the *Relations Switch* off.)

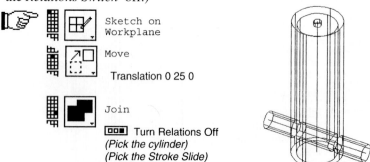

Sketch on Workplane

Move

Translation 0 25 0

Join

□□■ Turn Relations Off
(Pick the cylinder)
(Pick the Stroke Slide)

Sketch in Place on the bottom face of the Stroke Slide. Extrude and cut the existing inner diameter circle to trim the cylinder from the inside of the Stroke Slide tube. (Make sure the length of the arrow is long enough to trim past the cylinder, but not through the top.)

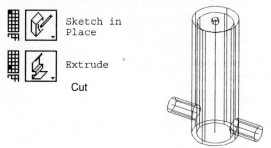

Sketch in Place

Extrude

Cut

Move the Stroke Slide 100 mm in the Y direction, so that it is positioned approximately mid-stroke. Name this part "Stroke Slide."

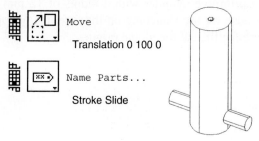

Move

Translation 0 100 0

Name Parts...

Stroke Slide

Get the Pivot Block. View the Stroke Slide and the Pivot Block from the front view. Sketch a shape as shown to design a part named Offset Link to connect the two.

Get

Pivot Block

Front View

Polyline

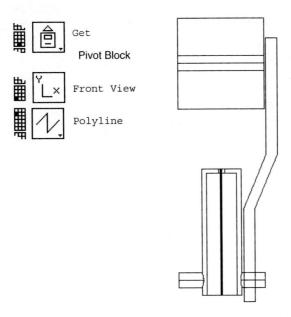

Extrude this shape 50 mm using the Thicken option to extrude in both directions, making a new part.

Isometric View

Extrude

New Part
Value **50**
Thicken

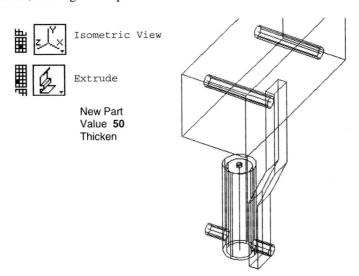

Sketch in place on the side of the Offset Link. View from the side. Draw a circle on this face with a radius of about 8 mm, using the *Focus* option to project the center hole location.

Sketch in Place

Right View

Center Edge

■□■ Focus
Pick circle center

Pick focused point
locate edge of circle

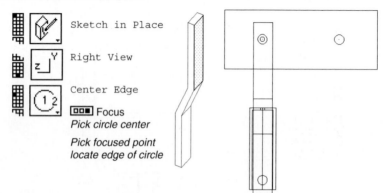

Cut this hole through the Offset Link.

Isometric View

Extrude

◆ Cut
(Use either a
depth great enough,
or use Thru All.)

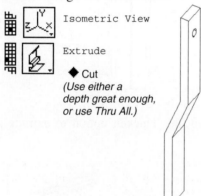

Use the Stroke Slide as a cutter to cut the hole in the lower end of the Offset Link. (Make sure you have named the Stroke Slide first, or it will be "absorbed" into the history of the Offset Link.)

Cut

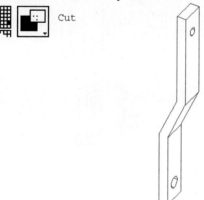

After the cut, the Stroke Slide used as the cutter will be put away in the bin.

Put away the Offset Link. Make a copy of it in the *Manage Bins* form, naming the copy "Offset Link - Copy." Get this copy out on the workbench and rotate it 180 degrees about the Y axis through the origin. Get the Offset Link on the workbench. (When you use the Assembly task, it is not necessary to make multiple copies of parts like this. The Assembly task can "instance" parts without duplicating the geometry. This will be covered in Chapter 6.)

Put Away
Offset Link

Manage Bins...
Offset Link
(Copy)
  Name: Offset Link - Copy
Offset Link - Copy *(select)*
Get (Icon)
Dismiss

Rotate
  Pivot point (Origin) ▢▢■
  About Y
  Angle 180

Tip: Try the Undo command here.

Get
  Offset Link

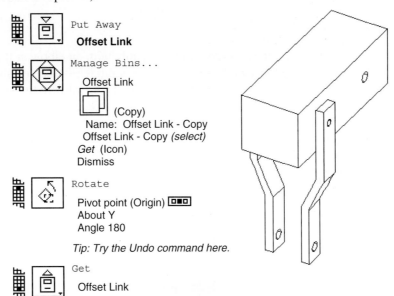

Now that all the parts are finished, get them all on the workbench so that you can see the entire pump mechanism. If any of your parts are out of place, use the *Measure* and *Move* commands to put them in the correct locations.

Get

Return the workplane to the global position.

Align

(Pick the workplane)
▢▢■ To Global

End the recording of your commands in the program file if you turned on create mode at the beginning of this workshop.

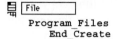
```
      File
    Program_Files
        End_Create
```

Save your model file.

```
      File
    Save
```

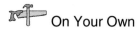 On Your Own

If you have time, design an output pipe, possibly with an elbow at the end to direct the water downward. (Hint: Revolve a circle 90 degrees about a line. Create an extra line to revolve about.) To finish off the well, it should be sealed at the base with a concrete pad so that run-off water can not get down into it. It should have a 2 meter diameter, and be 100 mm thick. Build a rectangular base to firmly hold the Casing and the Pivot Support tube in place. A suggested size is about 200 mm thick, and large enough to anchor the parts that protrude from it.

If you have access to libraries, check-in all of the pump parts into the Pump Parts library. This time, check them in with the library status *Keep for reference*.

When you are finished, save your model file and exit from I-DEAS.

```
      File
    Save

      File
    Exit
```

 Note:

You will need these parts for later workshops.

You may delete all your model files from this and the last workshop if your parts are checked into a library.

Chapter 3
Modifying Parts

Introduction

Engineers and designers may spend more time modifying parts than designing new ones. They often redesign parts for use in a new product. Even in brand new products, the design process is iterative with many changes made along the way. Conflicting design constraints need to be resolved such as interferences or weight and center of gravity (CG) requirements. Simulation results may dictate geometry or material changes. Manufacturing studies may suggest better design alternatives to save manufacturing and assembly costs. Marketing factors may dictate changes due to changing customer demands. These design changes will require part modifications. Because of this, it is important to understand the different methods for modifying parts. Also, knowing how parts can be modified will lead to good design practices to make parts easy to modify.

This chapter will first define what a part is, which is necessary before we describe how to modify them.

It is also necessary to have a good understanding of how to select the constituent elements of a part. The user interface is interwoven into the different operations of modifying parts, since the same commands are used in different ways to modify different aspects of a part, depending on what you select.

With an understanding of the basic terminology, we will next be able to define the various options available to modify parts and describe the program steps to make these changes.

What is a Part?

Five types of information define parts in the software, as illustrated below.

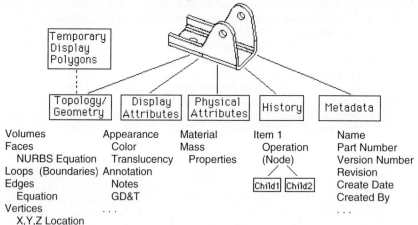

The five types of information are topology, display attributes, physical attributes, history tree, and data management metadata. Each of these five attributes will contribute to the discussion of how to modify parts, although the history tree is by far the most important.

Topology (Geometry Representation)

Historically, there have been three ways CAD systems stored part geometry in a computer: wireframe, surfaces, and solids. In a wireframe model, the equations of the boundary curves of the object are stored. With just a wireframe representation, it is impossible to check for three-dimensional interferences, since there is no representation for the surface between wires.

A surface model stores equations representing each of the surfaces of a part as well as the edge boundaries. The fancy animated 3D graphics generated for television "flying logos" only need surface modeling, since mass properties or other mechanical applications are not a concern. In a surface model, it may still be difficult to check for interferences, since we might not know for sure which is the inside or the outside of each surface. For the same reason, it will be difficult to compute mass inertial properties of such a part, especially if the surface model is not completely closed.

A solid part model must contain enough information to fully describe the boundaries, surfaces, and the topology of the part. If you cut a solid part, the result is another solid part, not an ambiguous collection of partial lines or surfaces.

I-DEAS part models contain a topology representation (T-Rep) that describes the part geometry. Topology defines how volumes of the solid are defined by faces, how faces are defined by loops, how loops are defined by edges, and how edges are connected at vertices. The geometry of these topological elements is defined by locations of vertices and equations for edges and surfaces. Each of these topological elements is tracked in the I-DEAS software by ID labels.

These labels can be turned on using the Display Filter icon. The *Display Filter* command also controls the topological and geometric information about a part which reaches the display screen. Part entities such as the key dimensions, centerlines of cylinders, and centerpoints of arcs can be turned on and off in a form. This form also controls the size of labels shown on the display.

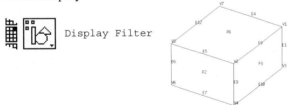

Display Filter

Features use these topological ID labels to keep track of relations between mating faces and edges. Understanding how relationships use these topological labels will give you a better understanding of what happens when you modify parts, and will help you resolve problems that may occur when you make dimensional changes that can cause large topological changes to the resulting part.

You may use the *Info* command to interrogate topological information about vertices, edges, and surfaces that make up a part. In general, surfaces are stored using NURBS (Non Uniform Rational B-Spline) surface equations. The Info command can give you information about the type of surface, such as planar, cylindrical, ruled, or other surface types.

 Info

Surfaces you see on the screen are only an approximation of the true NURBS surfaces used by the software to store equations defining the part faces. Most hardware graphic displays use faceted polygons to display surfaces. These polygons are not stored with the part; they are only created when needed for displays. You may note that as you zoom in and then redisplay the graphics, the display will be regenerated with smaller polygons.

The *Measure* icon subpanel is useful to investigate geometric information about a part. With these commands, you can measure distance and angles between the various entities of vertices, edges, and surfaces.

Measure

Distance

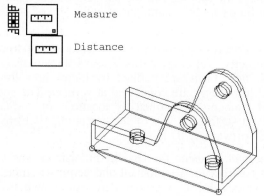

Display Attributes

Each surface of a part has display attributes such as color, glossiness, and transparency. Every feature you add uses a different color (in a programmed color wheel) to differentiate its surfaces from the rest of the part. The colors of these surfaces are not clearly shown in a line mode display, but are very apparent in a shaded display.

The *Appearance* command allows you to interrogate and modify these display attributes of a part, or individual selected surfaces.

 Appearance

You may change the color of the entire part to one color using this command. You may also want to change the color of individual surfaces to different colors to highlight certain features.

Physical Properties

Mass properties of a solid part are computed and listed with the *Properties* icon. The surface area, volume, and mass (for the given density), center of gravity location, and moments of inertia are listed. Moments of inertia are computed about the CG, the origin, and as principal moments of inertia at the CG. The principal axes will be displayed on the part.

   Properties

This command also will tell you what material is assigned to the part. Mass properties (other than surface area and volume) are not meaningful until a material with a density is assigned.

History Tree

Historically, commercial solid modeling programs have generally fallen into two categories: Boundary Representation (BREP) or Constructive Solid Geometry (CSG). BREP modelers store the results of operations (i.e., what exists after creation and manipulations), and CSG modelers store a history of the parts and operations required to construct the part. Solid modelers using the BREP method store the boundary topology as defined above. Older systems (including early versions of I-DEAS) used a faceted approximation instead of precise equations. The CSG method was slower, but had the advantage of greater flexibility of making changes to the part.

I-DEAS uses a hybrid data structure to store parts. Both BREP and CSG history information is stored. A "history" is stored for each part, which allows CSG type editing of the part history. This history contains the parts and the construction steps in creating the final part, plus the rules defining the design intent at each construction operation. For example, a hole drilled through a part could be located a fixed distance from an edge or a percentage distance along an edge. When the overall dimensions of the part are changed, the hole is relocated by the same rule.

Parts in I-DEAS are flexible or "rubberbandable," because the history of the part contains the design rules and relations defining how cutting features were oriented for the cut. The history of a part also contains parameters of features that were used to define them. The database also includes variational constraints such as parallel, perpendicular, or tangency constraints on profiles used as construction tools. Don't let these concepts scare you. All of this storage and calculation happens automatically as you create and manipulate parts. At this point, it is important to understand that all of this information is retained, so that you can make modifications based on the same input.

Part geometry can be modified by changing its dimensions or by changing the features, primitives, or sections which were used to create each feature of the part. A key concept is that the software remembers the "history" of a part containing the rules that were used to create it, so that changes can be made to any operation that was performed to create the final part.

The history tree is described by nodes and leaves, where the leaves are the features used and the node is the operation performed on the leaves, such as joining two features together. A typical history tree is graphically shown below.

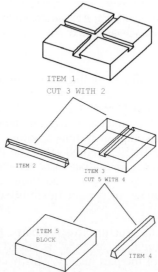

The history of a part can be graphically displayed using the *History Access* command.

 History Access

The graphical display will show the order of each step. This history tree gives information about modeling order and other information about the status of each step.

The history tree can also be listed using the Info command. The list that comes out will describe each item by the operation that was used to form that node of the history tree, and the items that were used in the operation.

It is important to understand the concept of the History Tree when you modify parts. When you select a feature of the part to modify, you may graphically pick an item from the part or you may use the menu options to pick a parent node (walk up in the history tree), or to pick a child of this node (walk down in the history tree).

Data Management Metadata

Metadata is data about data. In this case, metadata includes things like the part name and part number. It also includes information about when the part was created, who created it, and where it can be found. This information is stored by the data management system in the project file. Remember when you enter the software that it asks you for the name of a project? This project is the container for the project metadata.

The icons *Manage Bins* and *Manage Libraries* are the primary commands you will use to investigate metadata. The Manage Bins command can give information about parts and other items stored in the bins in your model file, where the *Manage Libraries* command can give information about items in any library.

  Manage Bins

In the form that is displayed, you can select items, and request different detailed information about the data management history of that item, such as when it was created, or the name of the parent item that contains it.

Modifying Parts

Each of the five aspects of the definition of a part described above can be modified separately.

Modifying the display attributes (color or glossiness); the physical attributes (attaching a different material); or the metadata (changing the part name or adding text) may be considered part modifications, but are not the main focus of this discussion, since these are relatively trivial operations. There are, however, some aspects of these that should be discussed.

Modifying Display Attributes

You may modify the display attributes of individual topological items (curves and surfaces), or the display attributes of the entire part. It is important to understand the picking operation and the graphical feedback it provides to know what you have selected.

In a team environment, you might use this as a modification strategy, for example to make the whole part one color, but change the color of surfaces of individual features that require further modification or definition to show up in a different color, so that team members can easily see them.

Modifying Physical Property Attributes

Physical attributes can be modified either by attaching a different material or by overriding the calculated mass properties with manually entered values. Many part designers may tend to ignore physical properties, but in a team environment, someone needs to have responsibility to enter the correct values at some point in the process.

Modifying Metadata

There are some limitations to changing the metadata assigned by the software. For example, you cannot change the version number assigned by the software or the creation date. Other data fields such as the name of a part can only be changed under certain conditions, such as when no other users have this part checked out. There are other metadata fields that are assigned by the user that you should learn to change. These include user-defined attributes which might be added when the project is configured, such as part cost or vendor names.

Another important field you can and should change is the *Change History* attribute. This is a text field intended to let you describe changes in words. Before you update a part from a library, you might check this attribute to see what the latest change was. If it was something minor that does not affect you, you might not bother to get the latest update. Because metadata is available even without getting the part, you can check this text field before getting the part. This field is only useful, however, if you develop a discipline to use it regularly to describe changes.

Direct Manipulation of Part Topology

Although direct manipulation of topology is not as common as history-based modification, to be described next, there are a few direct ways to manipulate topology that are useful.

One obvious way to directly manipulate topology is to delete a face of a part. This more often happens by mistake, when you meant to delete the whole part or a feature, but you selected a face by accident. This can be a useful technique, to allow you to manually delete a surface, and re-create it using the surfacing techniques discussed in Chapter 5.

It can also be useful to delete a face before shelling, although this is not necessary (or advisable) since the *Shell* command prompts you for faces to delete.

Direct surface shaping is another technique for manipulating topology. (In a technical sense, this is really modifying the geometry, not the topology, since the face ID numbers, edge ID numbers, and vertex ID numbers remain the same.) The *Drag Point* command allows you to pull on a point on a surface to directly shape it. This type of operation will be discussed in Chapter 5.

Although it is really just another part feature, partitioning has the effect of truly modifying the topology by converting the topology from "manifold" to "non-manifold" topology. This basically means that there are more or less surfaces than needed to bound a closed volume. Partitioning a solid with an interior surface may be used by Simulation users to make a model with different material properties or to provide control over the element meshing. This technique may be used in the design of a part to create an associative interior surface for the purpose of locating sections to be used in a loft, for example to define a complex cavity.

History-Based Part Modifications

There are several distinctly different ways to modify the part through its history tree. One category of changes is to change feature dimensions. This may only change the shape of a section defining a feature, but does not change the topology of the section or the history tree of the part. A slightly more radical change is to change the definition of the section defining a feature. The history tree may still be the same, but a feature of the whole part may have a radically different topology. A bigger change yet is to make a change that redefines the history tree of a part. For example, deleting a feature or inserting a new feature in the middle of the history tree both change the history tree definition. These history-based changes will be described separately, since they each have important uses.

Selecting a Feature

Before discussing modifications to features, it is important to have a good understanding of the methods for selecting an individual feature, as opposed to selecting a face or the part itself. There are at least four different ways to select a feature.

1. Select it graphically. If you can see a surface that uniquely belongs to a feature you may click on it until the feature is selected. It is important to recognize the graphical feedback that tells you when the feature is selected, since the number of clicks is different depending on whether you are pre-selecting or post-selecting. The feature is selected when its surfaces are highlighted in yellow, and a dashed yellow box surrounds the feature in space. Notice also the text written to the list window.

2. Use the *History Access* icon or right mouse button option.

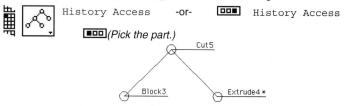

3. Use the right mouse button options *Parent* and *Child*.

4. Pick the feature by label.

Modifying Features

The general method for modifying features is to use the two icons– *Modify* and *Update*.

The *Modify* and *Update* commands permit you to make changes to existing parts. You can change dimensions of a part and modify size and placement of holes. You can also modify or replace extruded and revolved parts. When you update the part after defining the changes, the software uses the history tree to "replay" the steps to make the part, using the changes you have described with the *Modify* command.

When you select *Modify*, you have several options, depending on whether you have selected a feature (and what type of feature), or whether you have selected the part. The workshop at the end of this chapter will give you a chance to experiment with a simple part with features to see which options are available in each case, and what they do. Some of the important options available for features will be discussed next.

Modifying Feature Dimensions

With a feature selected, *Modify* options to change the feature dimensions include *Show Dimensions* and *Dimension Values*. The first option displays the dimensions on the part in 3D space, the second lists the dimensions in a table from where they can be modified.

Displayed dimensions can then be modified by again using the *Modify* command, and picking the dimensions directly. Modified dimensions will be shown in magenta to indicate that they are pending until you press Update.

When dimensions listed in the table are selected, the corresponding dimension on the part will highlight if it is visible. You may have to slide the form on the screen to see the dimensions behind it.

When the part is selected rather than a feature, the *Show Dimensions* and *Dimension Values* options show all the part dimensions.

Modifying Feature Parameters

The option *Feature Parameters* will present the form used to create the feature, and allow you to do things such as flip the direction of an extrusion or change an extrusion from a protrude to a cutout.

Modifying Feature Wireframe Sections

The Modeler task includes two kinds of wireframe geometry: curves and sections. Sections are a group of curves used to define a boundary. This boundary can be closed or open, and can contain holes. Sections are commonly used to create features by extruding or revolving. These sections and the curves defining them are stored with each feature. A useful concept is that of a "sketch pad." Each feature of a part contains its own sketch pad, which you can display and modify by using the Modify command with the Wireframe or Quick Wireframe options.

With the wireframe displayed, you can then modify curves and constraints defining the wireframe section, or modify the section by adding or deleting curves. To add or delete constraints, you use the *Constrain* and *Dimension* icons, which will be covered in the next chapter. You can also modify sections by dragging or modifying dimensions.

To add or delete curves to the section that traces around the boundary, first create any new wireframe curves you want to include. Then use the *Modify* command and pick the section. It may not seem obvious at first which you are picking, the underlying wireframe curve or the section which uses the wireframe. The software will select the section first, and prompt you to accept this selection or pick again. Next, the software prompts you to pick curves to add or delete from the section, just like when you first created it.

You may alternately delete the entire section tracing around the boundary and re-create it when you update the part, but it is normally recommended that you modify the section instead, to maintain associativity to later features that may be related to edges of this feature.

Depending on the version of the software you are using, there are two different options to display the wireframe, called "Wireframe" and "Quick Wireframe." One method "turns back the clock" to show you what the part looked like when the feature was sketched. This may take time to evaluate. The other method just shows the wireframe section in 3D space with the existing part. In most cases, the two methods will work the same. The only difference will be in the way you can focus on part geometry. The History Access icon also gives you the wireframe option.

Renaming Features

Another option is to rename the selected feature. This will allow you to name features things like "Mounting Holes" instead of "Extrude9."

 Naming features and dimensions can help you or someone else understand your part when you need to modify it.

Suppressing and Deleting Features

Selected features can be suppressed or deleted. These are not the same thing. If you suppress it, you can unsuppress it. If you delete it, it is gone. Suppressed features are displayed in the history tree with dashed lines.

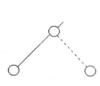 Suppressed features may be used in a case where a part is used in different applications with slight modifications. A feature could be suppressed for one application but left unsuppressed for another. This would allow you to use only one part model for both cases. Another common use for feature suppression is to remove small details for simulation or manufacturing.

Inserting Features Using Rollback

An option on the *History Access* form is *Rollback*. This option is allowed when a node (not a leaf) of the history tree is selected. This option will roll back the part to this level of the history tree, and temporarily set aside the steps after this point, so that you can add additional steps in the history tree.

Extracting and Reordering Features

Other options are to extract features, for example if you want to insert them into a different part. This is an effective way to reorder features.

Cases where you want to control the order of features include– where features overlap each other; to keep features that you modify often near the top of the history tree; to keep features that you don't change often near the bottom of the history tree; and to keep features containing reference geometry near the bottom of the tree.

Adding and Deleting Edges to Fillets

When a fillet feature is selected (actually when the "fillet info" leaf is selected), there is an option to add or delete fillet edges. This may be required when large topology changes occur, and the same surfaces no longer intersect where the fillet is to be placed. If the fillet gives warnings when you replay the part, you may need to use the option to delete unfound edges.

Adding and Deleting Feature Relations

The presence of relations, either sketch-in-place relations or relations created when the *Relation Switch* is turned on when performing a *Cut* or *Join*, is indicated by an "*" at the end of the feature name. One of the *Modify* options allows you to delete relations. The way to create relations is to create dimensions to points in the plane (such as by using *Focus* to project a point).

Minor Operations

Most operations you perform will add a node with two children to the history tree. An exception to this is minor operations such as changing the color of surfaces, adding reference geometry, or deleting surfaces from the part. These operations are graphically flagged in the history tree by a vertical line on the node as shown below on the left. If you double click on this node, it will expand to show the minor operations as triangles as shown below on the right. With the display expanded, you may select this operation and delete it.

History Tree Diagnostics

The history tree display gives useful diagnostic information about the status of the part creation, using color and line style.

The lines connecting nodes and leaves of the tree are normally green. If the lines are drawn in red or yellow, it means there were errors or warnings on that step. Just by looking at the history tree, you can see some diagnostic information. When you see red or yellow lines, you can select those features and modify them to correct problems.

Another visual cue in the history tree mentioned above is the "*" on the end of feature names that have relations to the rest of the part.

Before a design is released, parts should be checked for errors in the history tree. The history tree can also be used to display one feature at a time to make sure that every feature is correctly dimensioned and constrained, and that there are no features that have gross errors left in the tree, such as features cut in the wrong direction.

Documenting Part History

When the design of a part is complete, it is a good practice to give features meaningful names to make parts easier to modify. A practice that some users follow is to actually create a drawing of the history tree to document the part construction methods. There is a menu command *History Part* to construct the history tree graphically using wireframe lines and circles so that it can be plotted like other geometry.

Summary

This chapter covered a definition of what a part contains in order to define the different ways to modify this information. We then looked at several different ways to modify the definition of parts. The following workshop will demonstrate some of these important concepts.

☼ PARTING_SUGGESTIONS

-Save before you attempt radical modifications. To go back to the model file as it was at the last save, use the command *File, Open* or Control-Z. (Terminate any command before typing Control-Z.)

-Make sure you watch the graphical feedback so that you know the difference between selecting a surface, a feature, or the whole part.

-Use good modeling practice to make parts easy to modify.

-Rename features to make the part history self-documenting.

-When you modify features, you usually select the leaf, not the construction node to modify things like the feature parameters, wireframe, or feature dimensions. Select the node to roll back to insert a feature or to rename the node.

-Use metadata *Change History* to describe why you are making changes to each version.

-Use *History Access* to find mistakes in your part history.

-Someone should be responsible for checking the history of final parts before they are released. Are all history steps valid with no errors? Is the wireframe section of each feature fully constrained? (More on this in the next chapter.)

-Use *Help, On Context* to learn more about the specific operation of commands.

Where To Go For More Information:

For more information on the topics in this chapter, see the following.

Help, Help Library   or   *Help, Help Library, Bookshelf*
        Design User's Guide
            Part Modification

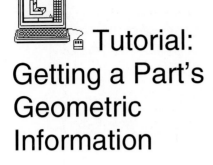

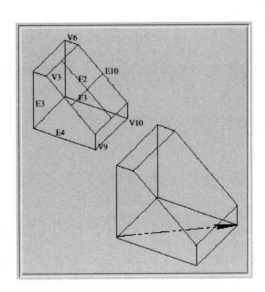

Tutorial: Getting a Part's Geometric Information

This tutorial will give you practice selecting part geometry and features, skills required to make part modifications.

Before you start:

1. You should know how to create parts and add features.

2. You should have a basic understanding of features, and how they are recorded in the part history tree.

What you should learn:

1. How to select part geometry and features using several different techniques.

2. How to recognize graphical feedback when selecting.

Tutorial Instructions.

Start I-DEAS with a new model file.

Follow the instructions in the tutorials below from the online Help Library.

Help Library, Tutorilas,
Design, Design Part Modeling,
Fundamentals,
Selecting Entities
Getting a Part's Geometric Information

Tutorial:

Modifying Parts

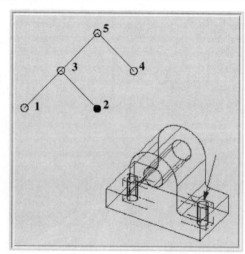

This tutorial demonstrates how to modify part features and dimensions.

 Before you start:

1. You should be able to select geometry and features, and recognize the graphical feedback when they are selected.

 What you should learn:

1. How to modify dimensions and parameters of features.

2. How to modify feature wireframe and section geometry.

3. How to insert and suppress features.

 Tutorial Instructions.

Select the first tutorial shown. After completing this tutorial, continue with the second tutorial.

Help Library, Tutorilas,
*   Design, Design Part Modeling,*
*   Fundamentals,*
*   Modifying Features*
*   Using History Access to Modify Features*

Try building the "On Your Own" part at the end. You should be able to change the first part into the second two by modifying feature parameters.

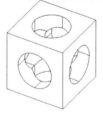 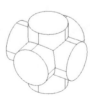

Workshop 3: Modify Parts

This workshop will help you review the concept of the history tree and use various methods for modifying parts. You will modify one simple part from the last workshop where you designed the parts for a hand pump.

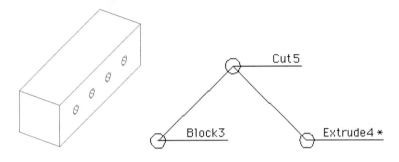

Before you start:

1. You should know how to create simple parts and add features.

2. You should know how to manage parts on the workbench and in the bins.

3. You should have completed at least Workshop 2A from the last chapter, where you put the pump parts into a library named "Pump Parts."

> **If you do not have this library, you can either:**
>
> 1. Run the program file Workshop2a.prg to re-create the parts. Or–
>
> 2. Go back and perform at least the first few pages of Workshop 2A to create the part "Pivot Block" in a new model file.

After you're done, you should be able to

1. Select part features using the history tree.

2. Modify feature parameters and wireframe geometry of part features.

Workshop Instructions:

Start up the Design application, the Modeler task with a new model file. (If you are already in I-DEAS, you can use the command File, Open instead.)

```
$ ideas <RETURN>
```

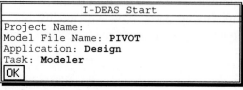

An idea is proposed to add an extra set of holes as a spare. When the holes wear out, the Pivot Block can be turned around to double its life.

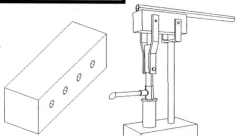

Get a copy of the Pivot Block from the library (or from a universal file) and get this copy out on the workbench.

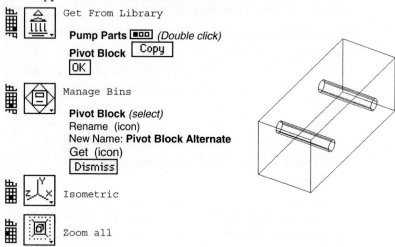

Get From Library

Pump Parts ▣▢▢ *(Double click)*
Pivot Block `Copy`
`OK`

Manage Bins

Pivot Block *(select)*
Rename (icon)
New Name: **Pivot Block Alternate**
Get (icon)
`Dismiss`

Isometric

Zoom all

 Note: Getting copies of parts from the library is usually bad data management practice when working with a team. You should use checkout if you will change parts, or reference them out if you don't intend to make changes. In this example, we will study a change, but won't check it back in, so we used a copy. Your company may have specific data management policies to follow.

Before you do anything, check to make sure your Pivot Block was constructed the same way as the one shown. The two holes should be cut as one feature, not two. Look at the history tree of the Pivot Block.

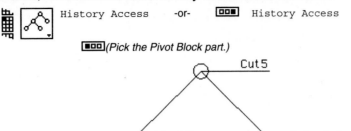

The history tree of your Pivot Block Alternate should have one node with two leaves. If your part has more steps than that, even if the final part looks right, you are wasting file space. Although not a problem yet with this simple part, when you model complicated parts, this type of inefficiency will add up.

> If your history tree does not look like the one shown above, you have done something wrong back in Workshop 2A. A typical problem is cutting the holes one at a time instead of defining the section to include both circles. Delete the Pivot Block part, then turn back to the beginning pages of Workshop 2A. Using the same model file you are in now, re-create the Pivot Block part as described in Workshop 2A. After you complete those steps, start over at the top of this page.
>
> ————————————————

 Modify the depth (Z) dimension from 400 to 500 mm.

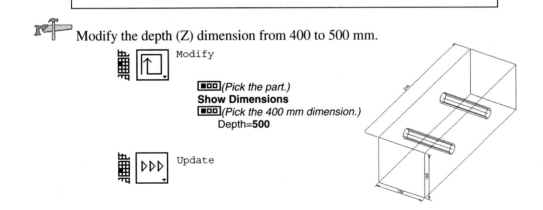

Pre-select the extrusion feature. Dashed boxes should then outline the two extruded holes. With this extrusion feature highlighted, select the icon command *Modify*. From the *Modify* options, select *Wireframe*.

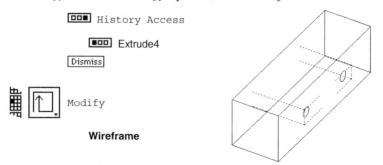

▣▢▢ History Access

▪▢▢ Extrude4

Dismiss

Modify

Wireframe

Add two more circles on the same horizontal line, each with a radius of about 15 mm.

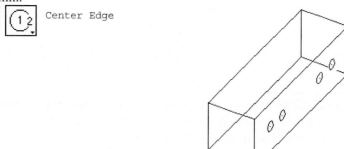

Center Edge

All four circles should have a diameter dimension.

Modify the value of the first circle to be a 30mm diameter. Modify the name of this dimension to be "Hole1." Modify the other three dimensions to *Match* the first. (Brackets around the dimension value indicate that it is defined by an equation.)

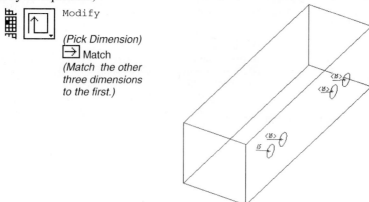

Modify

(Pick Dimension)
→ Match
(Match the other three dimensions to the first.)

Add dimensional constraints from the ends of the block to the outer two holes as shown below. Modify one of them to have the name "End_Distance," and the value 100 mm. Modify the other one to match.

Dimension

Modify

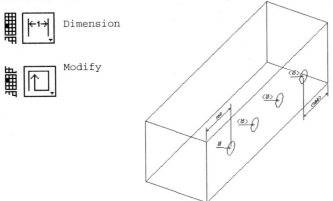

Add dimensional constraints between the first and third, and the second and the fourth hole as shown. Modify one to have the name "Pivot_Radius" and a value of 200 mm. Modify the other to match it.

Dimension

Modify

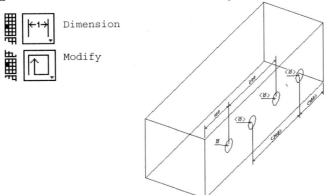

Modify the <u>section</u> that defines the extrusion to include the two new circles. The section is a "tracing" over the wireframe circles to define the section of the wireframe to extrude, it is not the circles themselves.

Modify

Pick entity to modify
(Pick highlighted circle section.)

Pick curve/section to add
*(Pick wireframe circles not
part of original section.)*

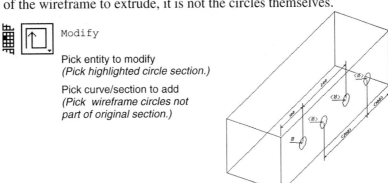

Update the Pivot Block twice. The first *Update* will update the extrusion feature. The second *Update* will update the part.

 Update

Update

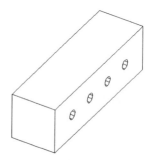

Note the process we followed in the design of this part– we first designed the shape we wanted, and then we sized the final dimensions. Not only is this a fast way to design parts, but it will make it easy to make changes to the dimensions that drive the shape now containing our "design intent."

Try modifying this part now. Use the *Modify* command; select the part; use the option *Show Dimensions* to show the key driving dimensions. Modify any dimensions you want, and then update the part. *(Since this part is a copy from the library, you don't have to worry about changing the original one.)*

Modify

▣▢▢ *(Pick part)*
Show Dimensions

Update

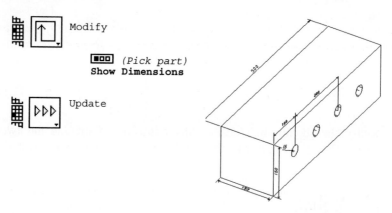

Part Modeling Strategy:

Display the history tree again after adding the two new holes. You will see that the number of steps in the history tree is still only two.

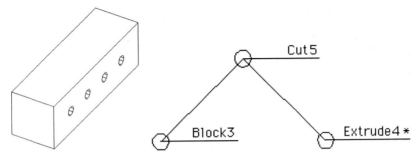

This shows that it is better to modify a feature than to add new features. This part could have been constructed with a total of five steps: one to create the block, and one step for each hole. This would take longer to create the part and take longer to modify it because the software has to run through more steps. It would also take more file space. More importantly, it would be harder for someone else to take your part and modify it. They would have to spend time to understand why there are five steps, what each of them does, and how they are related.

Can you see a better way yet to build this Pivot Block part? (With fewer steps.) If you think about it, you should be able to see that you can build it with only one step by extruding a rectangle with four circles inside. What would the history tree look like in this case?

In Workshop 2B, you used the Stroke Slide to cut a hole in the Offset Link. Get a copy of the Offset Link out of the Pump Parts library and display its history tree. Why is the history tree so complicated?

This illustrates that it is not necessarily good practice to use one part to cut another, since the entire history tree of the cutting part is inherited into the resulting part. It would be more efficient to extract just the necessary features from the cutting part to make the cut.

Some of the general rules you should use to decide how to model a part are:

1. Try to use the minimum number of steps to build the part.
2. Try to include as much detail in the wireframe of each feature as possible, rather than using multiple steps.

 On Your Own

Experiment with the different methods of selecting features. Practice selecting both the block feature and the hole feature using the following methods:

1. Graphically click multiple times on a surface of the feature until it is highlighted as selected.

2. Use the *History Access* icon.

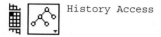 History Access

3. Use the right mouse button option *History Access*.

[□□■]   History Access

4. Pick the part, then pick *Child*, to walk down the hierarchy.

[■□□]   *(Pick part)*

[□□■]   Child

Between the above steps, you might want to deselect the last feature.

[□□■]   Deselect All

First with the part selected, then with each of the features (the block and the extruded holes) selected, experiment with the different modify options below.

    Which options are available when the part is selected?
    Which options are available when the primitive block is selected?
    Which options are available when the extruded feature is selected?

Modify Option	Part	Primitive	Extrude
Wireframe and Quick Wireframe			
Feature Parameters			
Show Dimensions			
Dimension Values			
Rename Feature (Try renaming Extrude4 to "Holes.")			

Save your model file and Exit when you are finished.

 ...

Chapter 4
Constraints & Constraint Networks

Design Intent

The solid model stores more than just the final geometry– it stores the "design intent" rules that govern what will happen when geometry changes. This makes it easy to make flexible design changes to I-DEAS part models. These rules are described as "constraints" on wireframe geometry used to extrude and revolve parts, and as "relations" when parts are cut, joined, or intersected. Part dimensions and constraints can also contain user-written equations.

These relations and constraints allow a design philosophy of "shape then size." Parts can be created quickly, and then constrained later to define specific dimensions or to add geometric design intent rules.

This chapter will first discuss constraints on wireframe geometry, and then relations on cutting operations.

Variational Geometry

The downstream flexibility of parts created from extruded or revolved sections in I-DEAS comes from the concept of variational geometry. A key to this system is what we call constraints, which are created when you sketch. Constraints look like dimensions when they are first applied, but they are more than that. They are a set of geometric rules that convey "design intent." If one dimension of a section is changed, how do we want the other points and curves to follow? This is determined by the constraints applied.

This concept is best described by a simple example. Start with a simple 4-sided polygon, defined by 4 points, 1 through 4, as shown below.

What do we want to happen if point 3 is translated to the right? Two possible solutions are shown:

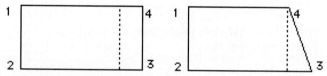

Although the original shape is shown as a rectangle, the person who drew it may or may not have had that in mind.

A way to solve this ambiguity is to constrain the right hand end so that it must remain vertical. What is happening in the software when this constraint is applied can best be described in terms of simultaneous equations. In a two dimensional plane, the four-sided polygon shown has eight degrees of freedom (DOF), since the location of each corner is described by an X and a Y location. To constrain the right end to remain vertical, a simple equation can be written:

$$X_3 = X_4$$

When this equation is applied, the number of DOF in the section is reduced by one. Other equations could be written to enforce other geometric constraints. For example, to constrain the top and bottom lines to be horizontal, the equations could be written:

$$Y_1 = Y_4$$
$$Y_2 = Y_3$$

Applying a dimensional constraint, such as to place a vertical dimension between points 1 and 2 to some value, say 100 for this example, is the same as writing an equation:

$$Y_1 = 100.0 + Y_2$$

Other equations can be written to express other geometric relations such as perpendicular and parallel. The equation to ensure that an arc remains tangent to a line may be more complicated, but the concept is the same. As we apply geometric constraints, the software is writing a system of equations that will be solved to determine the unknowns if any value is changed.

The beauty of this approach is that constraints can be added or removed in any order. This fits well in a design environment, since you may not initially know what constraints you have in mind as a design is sketched. Another reason this approach works well for conceptual design is that exact numbers are not needed as the design is initially sketched. As the design is refined, constraints are added or modified.

How Many Constraints do You Need?

Some mathematically-inclined readers may have a question at this point about how many constraints can be applied to a section. Normally, to solve a set of simultaneous equations, the number of equations must equal the number of unknowns. In I-DEAS, the number of DOF in the section is the upper limit to the number of constraints that can be applied. Once this many constraints have been applied, the section is fully constrained. The software will prevent you from applying too many constraints.

What happens if fewer than this number are applied, and a change is made to a value? There could be more than one possible solution. If you were using a general-purpose symbolic math package, it might not let you try to solve a set of equations where there were more unknowns than there were equations. The solver used to solve variational geometry is not just a simple simultaneous-equation solver. It looks at the geometry, and attempts to make the minimum change if more than one solution is possible.

Notice also that the software automatically creates most of the constraints you need. For example, while creating lines, if the Dynamic Navigator detects that you are creating a line vertical, horizontal, perpendicular, or parallel, it automatically creates these constraints on the line as it is created. The function of the right hand column of toggle switches titled *Constrain* on the *Navigator Options* form shown on page 50 is to allow you to turn automatic creation of constraints off, if you wish.

Tip: Don't inadvertently create constraints by locating points at the midpoint of lines, or aligned perpendicular or parallel if it is not your design intent to permanently constrain geometry that way.

Uses of Variational Geometry

The tools available with variational geometry can be used in different ways. For example, rather than creating a section to exact dimensions as you would in a drafting system, you could sketch the section approximately on the workplane, and then add variational constraints to "drive" the section into the configuration you want.

A snap grid is available as another way to locate points, but this seems more useful for demonstration than in practice. Using variational geometry makes using a grid unnecessary.

Variational geometry can also be used to study kinematic linkages. Constraining a point to point dimension on a line constrains the line to have a fixed length, like a linkage of a mechanism. Angular dimensional constraints can be applied, and modified to "drive" the mechanism to see its allowable configurations. This will be demonstrated in the workshop to follow.

When a constrained section is used to create a part by revolving or extruding, the resulting part will also contain the constraining "rules." Constraints which are applied to the section will be displayed as dimensions on the solid part when you pick the extrude or revolve feature used to create the part. Like other dimensions, they can be picked and modified to change the part.

Tolerance Analysis is another use for variational geometry. Variational geometry is the tool-kit that the Tolerance Analysis task uses to study what will happen as dimensions and tolerances are changed. This topic will be discussed in Chapter 7.

Creating Constraints

Constraints can be created in the Modeler task on wireframe or sections using the Dimension icon to create dimensions or the *Constrain & Dimension* icon to create dimensions and other types of geometric constraints.

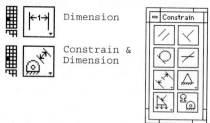

The *Constrain & Dimension* icon will bring up the *Constrain* icon panel, which contains icons for creating geometric constraints and dimensional constraints, and for checking the constraints that have been applied.

The constraints of parallel, perpendicular, tangent, collinear, and coincident are called geometric constraints. They enforce relationships between different pieces of geometry by internally writing equations that force the given geometric relationships to be true.

Dimensional constraints are equations where the given value is set equal to a constant. Dimensional constants can later be modified using the Modify command to set exact values, or the *Drag* command, to be discussed below.

The *Dimension* icon in the main icon panel will create linear, radial, or angular dimensions, depending on the curves picked. The icons inside the *Constrain* icon panel each create one of these types of dimensions separately.

Checking Constraints

While adding or deleting constraints, you will need to be able to understand what constraints have already been applied. The lower left corner of the Constrain panel shown on the previous page contains icons to interrogate which freedoms are constrained, and which are still free to move.

The *Show Free* icon can be used to tell you the status of curves. If you select this icon and then pick a curve or multiple curves (using shift-click), the program will tell you if each curve is unconstrained, partially constrained, or fully constrained. If a curve is fully constrained, it will be drawn in blue. If it is unconstrained, it will be left in green. If it is partially constrained, it will be drawn in yellow, and arrows will show the free directions. This icon can also be used on an entire section if you have created a section out of wireframe curves (using the Build Section command). In this case, the result will show the freedoms for every curve in the section, using the same blue and yellow colors to show fully and partially constrained areas of the section.

The *Show Constraints* icon is best used on one curve at a time to show how this curve is related by constraints to other curves. For example, this command will indicate that a line has been made parallel to a line on the opposite side of a section.

Drag

Another powerful tool in intuitively understanding the constraint network is to use the *Drag* command to dynamically shape a set of constrained wireframe geometry or a section built onto the constrained geometry. This command allows you to grab a curve or a dimension and dynamically change it in a rubberband mode, where the constraint equation network is being re-solved for each new position. This tool is primarily used to dynamically shape wireframe geometry and sections, but because it is continually re-solving the constraints, it is also very useful to interactively see the effects of the constraints in motion.

When you pick an unconstrained line or an arc with this command, you can move this curve in two directions with the mouse. All other curves constrained to this curve will move also, obeying all the constraints that exist. When you pick on a dimension (linear, angular, or radial), an "odometer" will be shown with the value of this dimension. You can dynamically change the value by moving the mouse up and down or left and right.

There are three different ways the *Drag* command can be used when modifying dimensions. If you pick either arrow head, you can increase the dimension in that direction. If you pick a witness line, you can attach it to a different point. If you pick the dimension text, you can reposition its location.

 It is easier to solve constraint problems when there are too few constraints than when there are too many. To understand constraints, a useful exercise is to delete all constraints, and add them one by one until the section is completely constrained and turns blue.

 3D Part VGX

A unique feature in I-DEAS is variational features. The *Constrain*, *Drag*, and *Dimension* commands previously described in this chapter for 2D sketching now work with the 3D geometry of features. This capability applies only to certain feature types such as extrude features.

To use this capability, you must first turn it on using either *Options*, *Preferences*, *Modeler/Assembly*, *3D Part VGX*, or the menus */Modify*, *Special Techs*, *Variational Extrude Sw* depending on your version.

In the *Constrain and Dimension* icon panel, the *Ground* icon can now be used to ground a face of an extrusion feature. In practice, you should ground three faces to define the datum of the part. Then, when you create face-to-face dimensions and modify or drag them, you will have defined that only one end can move.

Also in the *Constrain and Dimension* panel, the *Parallel*, *Perpendicular*, *Tangent*, and *Collinear* icons can be used to create 3D constraints between faces. Although this capability does not allow you to create any new types of geometry that you could not have done with 2D constraints, it means that you can directly create constraints between features without having to be concerned with which feature was created first in the history tree.

A very useful application of the variational feature capability is to use the *Show Free* icon to show which degrees of freedom are unconstrained in a feature. This capability is useful whether you have used 3D constraints or not. When you find unconstrained freedoms in a feature, you may use either technique to constrain it– adding the 2D constraints to the wireframe or adding 3D constraints to the faces of the feature.

Another powerful capability is to use the *Drag* icon on 3D faces. As you drag, constraints such as parallel, collinear, and tangent will be recognized.

Other preferences are *Drag for Extrude Creation* and *Drag for Revolve Creation*. With these switches turned on, you can dynamically drag the section to be extruded or revolved. Constraints as described above will be created if the *Constraint Recognition* preference is turned on. With this preference turned on, the extrude and revolve drag will snap to part faces and create VGX constraints.

Equations

When you modify a dimension, a form is displayed that shows the dimensional value as an equation such as D1=123.45. You may enter an equation rather than a constant in the dimension value field. For example, you may enter a simple expression such as D1=D2/2, where D2 is the label for another existing dimension. To find out the label of a dimension, you can use the Dynamic Navigator to show you. As you move the cursor over a dimension, the label name of the dimension will be displayed.

You can rename dimensions. For example, you could change the names to be something like "Flange_width," or "Total_height." When you move the cursor over the dimension, this name will be displayed.

A way to enter a simple equation when you want to match one dimension to another is to hold down the button ⊡ next to the dimension value field in the Modify Dimension form. Select "Match," which will prompt you to pick another dimension to match. The program will then automatically fill in the equation to make these two dimensions equal, such as "D2=D1." This could be used, for example, when you are sketching on a face and want to cut a pocket that is equally spaced from both sides of the part. This would build the constraint into the part such that if the part width was ever changed, the pocket would remain centered. Another use of the "Match" option is to create a simple match equation, which you can then edit. For example, if you want a dimension to be half of some other dimension, such as to keep a hole centered, you could edit the equation "D2=D1" to read "D2=D1/2."

Equations can also be created that are not directly connected to a dimension. When you select the Equations icon shown to the left, the program will ask you to pick a dimension and then will give you an equation form to enter any general equations. These equations are entered in the form of the "C" programming language. For example, you could write an equation:

```
if (total_height<200) then (Flange_width=50)
else (Flange_width=total_height/4)
```

In the field for the dimension values, you could then enter:

```
Flange_width ⊡ From Equation
```

Equation Units

All values are stored internally in SI units (meters) in the software. You must be careful about the units of terms in equations. For example, if you have an equation where you add a constant to a dimension, the software knows the units of the dimension, but it does not know the units of your constant. To avoid ambiguities, supply units to values in equations. Use the vertical bar symbol "|" around the unit names, such as |mm|, |in|, |m|, or |ft| as shown below.

```
D1 = D2 + 10|in| + 1|mm|
```

Relations

Relations are a way to capture design intent on *Cut*, *Join*, and *Intersect* commands, otherwise known as "Boolean" operations. If the *Relations* switch is off during the construction operation, no design intent relation is created. If the switch is on, the program will create a relation that is stored with the part history. If the dimensions of the part are changed in the future, the features will be positioned according to the relations that exist.

Creating relations on Boolean construction operations is basically a two-step process. The first step is to pick a face from the two parts to match up. The second step is to define the edge-to-edge relations within the plane where the faces are aligned. If the resulting part is modified later, the position of the two original parts will be repositioned to maintain the original construction relations.

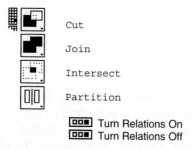

Cut

Join

Intersect

Partition

Turn Relations On
Turn Relations Off

The *Relations* switch should be on if you want the feature to be associative to part changes. If this switch is off when the feature is joined, it will not be repositioned if the resulting part is modified.

If the *Relations* switch is on, you do not need to manually position the parts first before selecting the Boolean command. The command will prompt you to define the relation, which will both position the parts and create the relationship. If the *Relations* switch is off, it is entirely up to you to position both parts in 3D space where they should be located for the Boolean construction operation, and they will remain in that location regardless of later changes in part geometry.

Parameterized Parts

Parameterized parts in the Modeler task can be stored in a part catalog or in a feature catalog to be used in construction operations.

To create parameterized parts and features, first create a generic part. Make sure that this part is constructed using constraints and relations so that the part can be easily modified by just changing dimensions. Another important step is to give the dimensions names like "Slot Width" instead of just "D14".

When the part is ready to create a parameterized part, use the icon *Modify Catalogs* to bring up the sub-panel shown.

 Modify
Catalogs

Use the icon in the upper left, *Parameters*, to define which of the parameters of the part are modifiable when the part is retrieved from the catalog. If you want to define tables of these parameters, use the icon in the upper right, *Family Table*. This lets you define standard configurations instead of modifying every individual parameter when you retrieve the part.

Check the final parameterized part into the Part Catalog or the Feature Catalog with the icon in the lower left, *Check In*.

When you get this feature or part out of the catalog later, you will be presented with a form to fill in the values of the variable parameters.

Note that to use parts or features from a catalog, you will use the icons shown on the previous page. The icon in the lower right of the *Modify Catalog* sub-panel is used to get the part or feature back to modify its definition, not just to use it.

Summary

This chapter has discussed how variational geometry, equations, and relations are used in the I-DEAS software to capture design intent. This discussion becomes most important as you begin to modify designs, which happens when you iterate to converge on an optimum design, especially when working concurrently with a team of other users.

Where To Go For More Information

For more information on the topics in this chapter, see the following:

Help, Help Library
*   Design User's Guide*
*   Part Design*
*   Part Creation*
*   Constraining Wireframe Geometry*
*   Constraint Shortcuts*
*   Creating Equational Constraints*
*   Part Construction*
*   Using Variational Extrude*

☼ PARTING_SUGGESTIONS

-Use the *Focus* option to project points or lines from other features on a part to the face you are sketching on. Create dimensions to these points to create associative relations.

- When sketching, be careful not to create constraints you do not intend to apply. (Use the control key to temporarily turn off the Navigator.)

-When adding constraints, first create constraints that do not change the geometry, such as grounds, dimensions, and fillets.

-Constrain the displacement of the section by grounding at least one point in the x and y directions. Also constrain the angle of the section by grounding a second point or the angle of a line.

-Don't over-ground the section. This may lead to a design that is inflexible and difficult to modify.

-Ground lines, then make other lines perpendicular or parallel to those lines, rather than using more ground constraints.

-Avoid large changes to dimensional constraints until the section is fully constrained.

-To maintain the direction of dimensions, use point-to-line or line-to-line. Point-to-point dimensions are allowed to reverse their directions.

-If undesirable results are obtained by the variational geometry solver, use the *Undo Last Process* command.

-If large changes in scale are required, use the *Modify, Scale* command instead of changing dimensions on the section.

-The *Automatic Wireframe Update* switch can be turned off in *Update Options* to take the solver out of interactive mode, allowing multiple changes to be made at once.

-To be used in a part, a section does NOT have to be fully constrained, but this is normally recommended before completing the design.

-The solver updates the geometry relative to the grounds that have been defined. If no grounds exist, normally the first entity created remains stationary and other geometry moves.

-Use the *Appearance* icon to autoscale dimension text size. Hint-Preselect one dimension, then pick *All* from the pop-up menu. Then pick the *Appearance* icon to autoscale all dimensions. Set this size as the default by pressing the button *Set As Default*.

-Turn on Variational Features with the menus: *Modify, Special Techniques, Variational Extrude Switch*.

-Use *Modify, Show Dimensions* to display 3D dimensions and constraints.

-Be sure to *Save* often, because it may be more difficult to delete improper 3D constraints.

-To delete 3D constraints, use *Filter, Constraints, Pick Only*.

Tutorial: Sketching and Constraining

These tutorials will give you practice with the important skills of sketching and constraining.

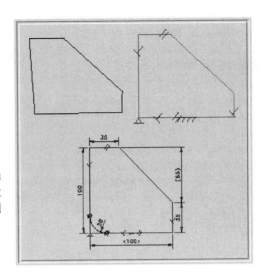

Before you start:

1. You should know how to create parts and manage them in the bin.
2. You should know how to select geometry.

What you should learn:

1. How to set Navigator options to control the automatic creation of constraints.
2. How to create, modify, and delete constraints.

3. How to find unconstrained free degrees of freedom.

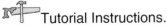Tutorial Instructions.

Start I-DEAS with a new model file.

Follow the instructions in the tutorials below.

Help Library, Tutorials,
Design, Design Part Modeling,
Fundamentals,
Sketching and Constraining
Dimensioning

Tutorial: Using Sections and Sketch Planes

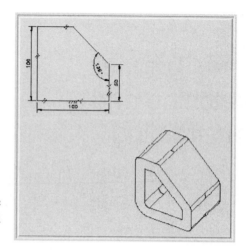

These tutorials demonstrate techniques of creating and modifying sections.

Before you start:

1. You should know how to sketch wireframe geometry and create constraints.

2. You should understand features, how they are recorded in the part history tree, and how to modify the section defining a feature.

What you should learn:

1. How to create and modify sections.

2. The concept of feature sketch pads.

3. How to move geometry between sketch pads.

Tutorial Instructions.

Continue using the model file from the previous tutorial page.

Follow the instructions in the following tutorials from the online Help Library.

Help Library, Tutorials
    Design, Design Part Modeling,
        Fundamentals,
            Building Sections
            Using Sketch Planes and Understanding Sketch Pads

Tutorial: Adding Features With Associativity

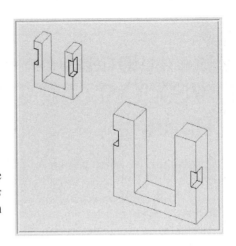

This tutorial demonstrates feature relationships starting with *Focus* and *Match*, and continuing with part equations.

Before you start:

1. You should know how to create parts and add features.

2. You should know how to sketch and add dimensions and constraints.

What you should learn:

1. How to create simple relationships between features using *Focus* and *Match* dimensions.

2. How to use equations to create relationships between features.

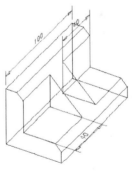

Tutorial Instructions.

Start I-DEAS with a new model file.

Follow the instructions in the tutorial below from the online Help Library.

Help Library, Tutorials,
Design, Design Part Modeling,
Fundamentals,
Adding Features with Associativity

Tutorial: 3D VGX Options on Parts

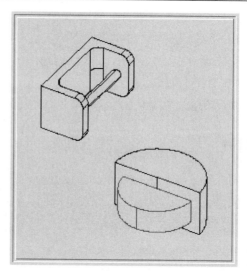

Constraints between features can be placed directly on part faces. In the newest release of I-DEAS, these constraints are created when using the option to drag the section to create extruded and revolved features.

Before you start:

1. You must have at least I-DEAS 8 to complete this tutorial.

2. You should know how to create features, and understand how to modify them using the part history tree.

What you should learn:

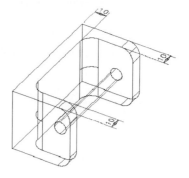

1. How to use drag to extrude and revolve features.

2. How to add and modify constraints on part vertices, edges, and faces.

Tutorial Instructions.

Start I-DEAS with a new model file.

Follow the instructions in the tutorial below from the online Help Library.

Help Library, Tutorials,
Design, Design Part Modeling,
Fundamentals,
3D VGX Options on Parts

Workshop 4:

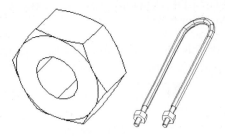

This workshop will give you
practice using constraints.

In this workshop you'll model a hex nut that can be sized by its key
dimensions. The first step is to create a hexagon that can be "driven" by
one dimension, the distance across the hexagon from one side to the other.

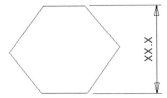

The starting six-sided shape has twelve degrees of freedom, two at each
corner, so twelve constraints will be needed to fully constrain the shape.
There are many ways to do this. One solution will be shown here. The
desired outcome is a fully constrained hexagon that can be modified by
only one dimension.

Although there are other ways to create a hex nut such as the *Polygon*
command, or by retrieving the completed geometry from the web site
MDCyberCad.com, it is a good exercise to practice creating constraints.

Before you start:

1. You should know how to create and modify parts.
2. You should know how to modify parts using the history tree.

After you're done, you should be able to:

1. Sketch properly constrained wireframe sections.
2. Create sections that can be sized by modifying a minimum number
   of driving dimensions.

Workshop Instructions:

Start I-DEAS with a new model file. Get into the Design application, the Modeler task.

For this workshop, turn off the Navigator options which automatically create dimensional constraints.

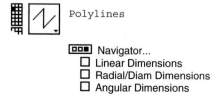

Start by sketching a six-sided polygon on the workplane. Be careful when you sketch this shape that you do not let any lines snap perpendicular to others. If you do get any perpendicular constraints, delete them.

Bring up the *Constrain* sub-panel, and ground one point. Make sure every side is constrained to be parallel to the opposite side.

If applying any constraint causes the shape to jump to an undesired configuration, use the Undo command to delete the last constraint.

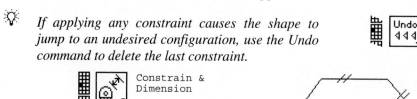

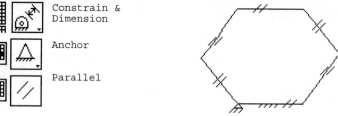

Notice the color of the wireframe lines while you are constraining or dimensioning. Lines in green are unconstrained, lines in yellow are partially constrained, and lines in blue are fully constrained.

Add three linear dimensions as shown, from <u>line-to-line</u>, not point-to-point. (Don't worry about the actual dimension values yet.)

`Dimension`

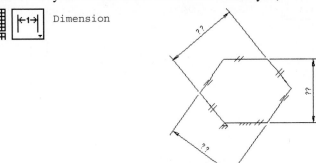

Modify the first dimension to give it the name "Head_Size." Modify the other two to match the first.

Modify

→ Match

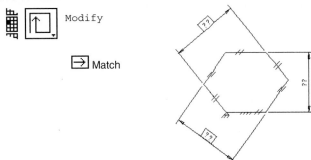

Dimension the half-width of the head size as shown. Pick a <u>line and a point</u> to create this dimension. Modify this dimension so that it equals Head_Size/2. (Use *Match*, then type in the rest of the equation.)

Modify

→ Match
Head_Size/<u>2</u>

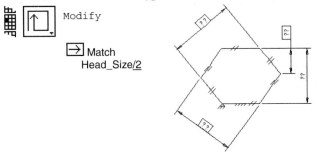

Add two angular constraints and modify their values to be 120 degrees.

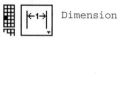 Dimension

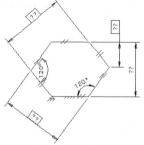

As you add the last angular dimension, all of the lines should turn blue, indicating that the shape is fully constrained. This means that a total of twelve constraints have been applied, since the original shape had twelve degrees of freedom.

To test the constraints, use the *Drag* command on the arrowhead of the longest vertical linear dimension. Modify this dimension to be 20 mm.

 Drag

(Pick the arrowhead from the linear dimension without the box around it.)

Modify

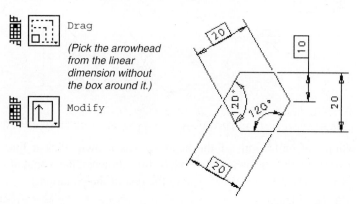

Extrude the section 10 mm.

Extrude

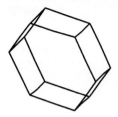

Sketch in place on one side of the hex and draw a circle. Constrain its diameter, and modify its name to be "Hole_Diam" and its value to be 10 mm.

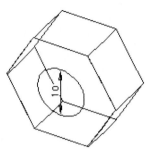

Extrude the circle, cutting through the part. Name this part "Hex_Nut."

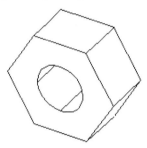

The threads on the nut usually are more detail than you want in a solid model, but one detail modification you may wish to make is to bevel the corners.

Sketch in place on the face of the nut, draw a centered circle with its diameter touching the sides. Delete the diameter dimension.

Constrain the circle to be tangent with one of the sides. (This will ensure that if the size of the hex nut changes, the circle will also change with it.)

Extrude this circle with a 60 degree draft angle, intersecting with the part.

 Extrude

Intersect
Draft Angle 60
Thru All

Repeat these steps for the other side as shown above.

Now that you have a "rubberbandable" part, you could use it to create any size hex nut by modifying a few dimensions.

On your own:

Redesign the Offset Link part of the pump mechanism using variational geometry. (Rather than modifying your part in the library, create it over from scratch.)

Start by sketching the shape shown on the right.

Add dimensional constraints so that the shape of the link can be "driven" by dimensions with the names: Length, Thickness, Offset, Upper_Length, and Lower_Length.

Extrude this shape 50 mm.

Add two holes that are constrained with radial dimensions named Upper_Radius and Lower_Radius. Constrain the locations of these holes so that they will remain centered at a fixed distance of 25 mm from each end.

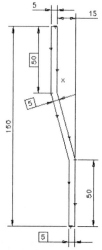

 Another interesting use of variational geometry is to model a mechanism. A variational geometry network can be constructed of our pump mechanism (as shown below), which allows "dragging" the handle angle to see the range of motion of the mechanism.

Hints:

Start with a workplane big enough to fit the sketch. Make the sketch of the parts at an angle as shown, so that the Dynamic Navigator does not add extra horizontal and vertical constraints that do not belong. The handle and the pivot block are represented by two lines that have been constrained to be collinear. (This happens automatically if you pick a point on a line.)

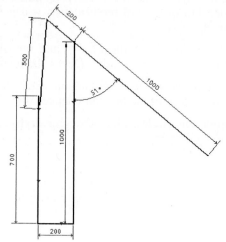

Chapter 5
Surfacing Techniques

Surfacing

So far in previous chapters we have described how to create parts from basic primitive shapes, extrusions, revolves, and construction operations. As you look around you, it is easy to find objects (such as molded plastic parts) with flowing sculptured surfaces. It would defy your imagination to figure out how to model them using the methods we have described so far. This chapter will introduce some of the tools available for creating features of more complex parts with sculpted surfaces.

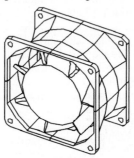

For this discussion, these techniques are divided into two categories–solids-based creation techniques and manual surface creation tools. The parts shown below will be created in the following workshop to illustrate the solids-based techniques of lofting and sweeping. These techniques create solid parts and features directly, much like the *Extrude* command, only using more sections in the definition.

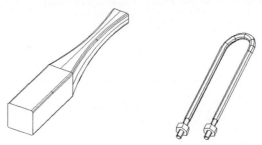

The second category of surfacing tools are manual techniques to create surfaces from wireframe geometry. These tools will require more steps to create solid parts from the bottom up, starting with wireframe geometry to create surfaces, and using surfaces to create solids as shown below.

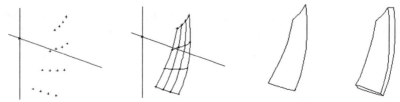

Solids-Based Surfacing Tools

The techniques of sweep and loft work very much like extrude and revolve. They use wireframe sections to directly create new solid parts or features protruded or cut from other parts. The major difference is that these commands use multiple sections to define the feature. A part such as the boat hull below is conveniently described by a series of 2D cross sections.

Like *Extrude* and *Revolve*, sections can be built first or created "on the fly" by the command that uses them. They can be thought of as just a grouping of curves. In most cases where the software asks for a section, you can either pick an existing section or pick the curves directly.

Sweeping

For swept parts, two different section definitions are required. One section will be "swept" along a path defined by another curve or section. The path can also be an edge of an existing part.

The path used for sweeping normally must have a continuous first derivative. This means that there cannot be any corners in the path. (In I-DEAS 8 or later, there is an option for providing a blend at the corner.) For example, the path on the left below is invalid for sweeping, the path on the right is OK. If your path is made from straight lines, you will need to fillet the intersections to create a smooth path for sweeping.

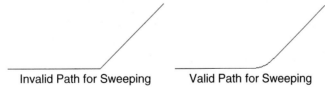

Invalid Path for Sweeping   Valid Path for Sweeping

More than one section can be supplied for the section to be swept along the path. The resulting part will gradually transition between these sections. Options exist to define the placement of the sections.

The section to be swept can be open or closed. If a closed section is swept into a part, the resulting part is a solid part. If an open section is used, the resulting part will be an open surface, not a closed, solid part. This surface can be turned into a solid part by several methods which will be described later.

Other options exist on the *Sweep* command. For more details on these options, read the online documentation using *Help, On Context*. For example, there is an option to orient the sections at a fixed angle instead of perpendicular to the path curve. Another option is to interpret the swept section not as it looks, but to use it as if it were a plot of the radius of a circular cross section. This option is called a "circular law."

There are two icons for sweep. The first is the *Sweep* command described above, and the other is called *Variational Sweep*. This second command sweeps a section controlled by variational constraints along a path, where points on the section can follow other rail curves.

Lofting

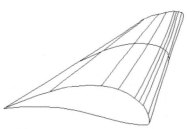

Lofted parts are defined by a number of sections positioned in 3D space. This is similar to sweeping, but a path curve is not used.

The sections to be lofted can be pre-existing sections defined with the *Build Section* icon, wireframe curves, or surfaces of existing parts. For example, you may want to loft a shape between faces of two existing parts.

When you pick the sections of the loft, it makes a difference which curve on the section you pick, and where on the curve you pick it. The end nearest to the pick location will be used as the starting point for the section. The direction from this end point toward the pick location defines the ordering of the curves around the section. It is important that all the sections used in the *Loft* command be picked in a consistent fashion, or you may get a twisted part that is not what you wanted, unless you are designing a screw or a twist drill. The first or last section can degenerate into a point to model blunt or pointed-end parts.

The number of curves should also be the same for each section. The program will handle sections with different numbers of curves, but you will have better control over how surfaces flow from one section to the next if they are divided into curves as consistent with each other as possible. The connecting lines between sections will be displayed and can be modified. Other lines can be added or modified to control how the surfaces will flow between profiles, but this is not necessary if you define each of the sections to have the same number of curves.

For detailed information on all of the options available with *Loft*, use *Help, On Context*. Also, see the tutorial "Creating Lofted Features."

Manual Surface Creation Tools

Although it is usually preferable to create parts using solids-based tools, there are applications where surfaces must be created and manipulated manually. For example, parts like the surface of the fan blade at the bottom of the page are not easily defined by flat cross sections. This fan blade is more easily defined by a mesh of points or curves in 3D space generated in cylindrical coordinates.

Other cases where direct surfacing may be desired is where the surfaces or wireframe curves are imported from some other source, such as the standard IGES file format.

Creating parts from surfaces is a multi-step process. First, the wireframe curves are created and manipulated. Variational Shape Design tools can be used to bend and style these curves into the desired shape. Next, these curves are used to define surfaces and/or solid parts using one of the methods that will be described below. One more step is necessary if you want to generate a solid part from the resulting open surface. This might be a shelling operation to give the surface thickness, or stitching multiple surfaces together to form a closed solid.

Surface Creation

Surfaces can be directly created using the commands previously discussed, including *Extrude*, *Revolve*, *Sweep*, and *Loft* by using open sections. For example, extruding a single line will create a planar surface. Surfaces can also be created from wireframe geometry using the methods "Mesh of Curves" or "Surface by Boundary." With either method it is necessary to define bounding curves first.

With *Mesh of Curves*, you might start with a grid of points as shown below on the left. These points could be used to define a grid of curves in two directions along the surface as shown on the right. The curves do not actually have to touch, because the program will average between them if they don't. Parts which converge to a point can be defined by picking the same end points for each column of points, or by selecting curves that converge on the same end points.

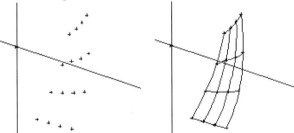

Surface by Boundary is an easy method to use when the surface is adequately defined by the boundary alone. This is often the case when the appearance of a part is defined more by the intersections or breaks in surfaces than by the exact curvature of the surface between the boundaries. To use this method, you could first create a wireframe representation of the surface intersections, then create surfaces inside each boundary.

Another typical use of the *Surface By Boundary* command is to manually delete a surface or a portion of a surface from an existing solid part, and then create a new surface to fill the hole defined by its boundary.

Curves on Surfaces

A typical surface operation is to trim or split surfaces at specific curves on the surface. An example of this is to project a company logo onto a surface. The common method to create curves on planar surfaces is the *Sketch in Place* command, however this does not work on curved surfaces. Curved surfaces require other methods to create curves on the surface.

The *Project* command can be used to project 2D wireframe letters or logos from a plane onto a curved surface. The curves could then be used to split the surface so that the area inside the logo is a different surface than outside, so that the surface inside the logo could be given a different color. Another possibility is to use the surface inside the trimmed boundary to create an offset surface to emboss the logo onto the parent surface.

Another example is to create the parting line of a mold for a part with complex surfaces. The *Silhouette* command can be used in this case to create curves on surfaces. This command creates a curve defining the silhouette as seen along a vector of the pull direction of the mold. The resulting curve could be used to split the underlying surface to create the parting line of the mold. Some operations will require that the surface be split into two surfaces along this line, such as to modify the draft angle on either side of the parting line, or to create a new surface to cut the mold block into a core and cavity.

Surface Intersection is another command where the result created is a curve that lies on surfaces. Again, the resulting curve could be used to trim the surfaces.

Surface Manipulation

After using the basic tools to create surfaces and to create curves on surfaces, surfaces often must be manipulated to make new surfaces. The commands *Trim*, *Split*, *Extend*, and *Merge* are commonly used to modify surfaces. Other options of the *Extrude* and *Revolve* commands are *Split* and *Partition*, which also can be used to divide surfaces and volumes of parts.

For more information on these surfacing commands, use *Help, On Context*.

Solids from Surfaces

An open surface can be turned into a solid part using the *Shell* command, giving the surface thickness.

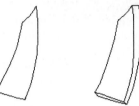

It is also possible to create a solid part from surfaces by stitching enough of them together to "sew up" a valid solid part. The *Stitch* command joins surfaces at common edges.

This technique may be used where one surface of a solid needs to be replaced. The particular face can be deleted, then a new surface created and joined back to the solid to close it up.

Surface Evaluation Tools

The Modeler task also has tools for dynamically evaluating surface quality parameters such as radius of curvature and surface normals of a surface. The lofted wing shape from a few pages back is shown below with a contour plot of its curvature. Another type of display is a "hedgehog" display. For information on this type of display, search for the term "hedgehog" in Help.

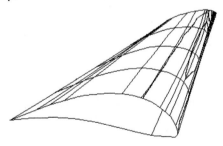

Open Part Modeling

Some thin-walled or stamped sheet metal parts may be most conveniently modeled as an open surface, stamped to apply features, and then shelled to supply the final thickness. When performing construction operations between open surface parts and closed parts, it is important to understand the importance of the "material side" in the construction operation. For example, if a part is cut by an open surface, material is removed on the side identified as the material side of the surface. The material side is defined using the *Material Side* command. It can be either side, both sides, or neither side (none).

Variational Shape Design

Curves in I-DEAS are mathematically defined internally as NURBS (Non Uniform Rational B-Splines). These curves are defined by control points that do not lie on the curves themselves. Rather than force the user to deal with the location and weights of these control points, I-DEAS lets the user shape curves in ways that are more intuitive to the user. I-DEAS tries to mimic the way things bend in nature, as this is more intuitive than the mathematics of control points. I-DEAS offers two different categories of shaping tools– direct shaping and energy-based shaping. Direct shaping allows you to directly push and pull on curves to manipulate them. Energy-based shaping "puts physics" into the curve shapes, and allows you to bend the curves with different types of forces.

The icon *Shape Design* brings up a sub-panel of icons used to manipulate curves. The first row of icons in the *Shape* sub-panel allows direct manipulation of curve geometry. Use the *Drag Point* icon in the upper left of this sub-panel to "grab" a curve at any point and dynamically move it in the plane or out of the plane. While you are moving the curve, options are available in the pop-up menu such as to move normal to the plane and to increase or decrease the range of influence on each side of the point being moved. Accept a final curve shape with the left mouse button. Cancel without making a change with the middle button.

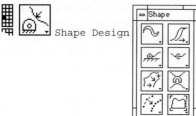

Shape Design

Icons in the second row apply constraints to curve ends, much like the variational geometry constraints applied to lines and arcs in the last chapter.

Icons in the third and fourth row in the sub-panel above perform energy-based and curvature shaping. These commands iteratively shape curves based on *Constraints*, *Forces* (*Push*, *Twist*, *Magnet*), and the *Prototype Geometry* applied to the curve. Prototype geometry is the natural free state the curve will tend toward in the absence of other forces, much as a stretched rubberband tends toward a straight line.

The *Function* command can be used to create a function of curvature or other property of a curve. This function can be edited directly to give very fine control over the curve to be modified. These functions can also be stored, to use for prototype geometry for other curves, to shape curves to look like other curves. These tools are useful for the design of critical surfaces such as airfoils.

Summary

This chapter introduced some of the surfacing and shaping tools. The basic methods are sweep and loft to create solid shapes from multiple sections. Other methods are also available to create surfaces directly from wireframe geometry, and manually stitch them together into solids.

Where To Go For More Information:

For more information on surfacing techniques, see:

Help, Help Library
  Design User's Guide
    Surfacing
      Surface Creation
      Surface Evaluation
      Surfacing Techniques and Examples

⚬ PARTING_SUGGESTIONS

Sculpted Parts:

- If it is possible to create a feature using simple techniques such as extrude with draft and applying fillets, it is usually preferable to do it this way.
- Use "bottoms-up" surface creation and stitching into solid parts only as a last resort, such as when forced to use geometry from external CAD systems. It is preferable to model parts using only construction operations as shown in earlier chapters.

Lofted Parts:

-Avoid "twisting" lofted parts caused by starting sections at different locations. (The pick location defines the starting point and direction around the section.)
-Although you can modify the connecting lines between sections, it will be easier to keep track of the number of curves used on each section when you create them, and make sure each section in a loft has the same number of curves.

Swept Parts:

-The path curve must have a continuous slope. (No sharp corners.)

Mesh of Curves:

-Although not required, it is recommended that curves actually intersect.
-Use smooth curves without corners.
-Try to use uniformly spaced intersections, with intersection angles as close as possible to right angles.
-Extra cross section curves may have to be added to capture peaks, rapidly changing areas. Flat areas of the surface may also need intermediate curves to control the "oscillation" of the spline surface equations.
-Use as few curves as possible to describe the surface.

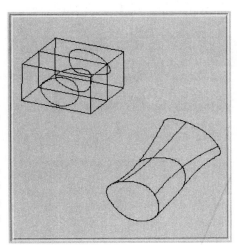

Tutorial: Creating Lofted Features

This tutorial teaches the basics of lofting features. It also demonstrates some techniques for creating associative planes for sketching.

Before you start:

1. You should know how to sketch wireframe geometry and create constraints.
2. You should know how to add features using commands like *Extrude* and *Revolve*, and be able to modify them using the history tree.

What you should learn:

1. How to create lofted features.
2. How to properly pick the sections for lofting.
3. How to use part faces and partitions as sketch planes.

Tutorial Instructions.

Start I-DEAS with a new model file.

Follow the instructions in the following tutorial found in the online Help Library.

Help Library, Tutorials,
Design, Surfacing,
Fundamentals,
Creating Lofted Features

At the end of this tutorial, try lofting the faucet shape in the "On Your Own" section.

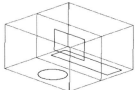 

Tutorial: Creating Swept Features

This tutorial demonstrates techniques where you can use swept features.

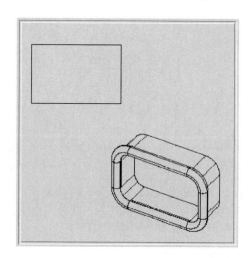

Before you start:

1. You should know how to sketch wireframe geometry and create constraints.

2. You should understand how features are recorded in the part history tree, and how to modify the wireframe section defining a feature.

What you should learn:

1. How to create the wireframe geometry defining a swept feature.

2. How to create a swept feature using wireframe curves and part edges.

Tutorial Instructions.

Open the tutorial listed below.

Help Library, Tutorials,
Design, Surfacing,
Fundamentals,
Creating Swept Features

An example of a part using *Sweep* is a paper clip, modeled in this tutorial.

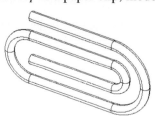

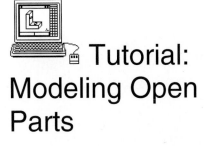

Tutorial: Modeling Open Parts

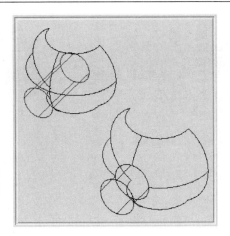

This tutorial demonstrates the basic techniques of creating and manipulating open surface parts.

Before you start:

1. You should know how to use *Cut* and *Join* with solid parts.

2. You should know how to sketch and add dimensions and constraints.

What you should learn:

1. How to create open surface parts.

2. How to set the material side.

3. How to use *Cut* and *Join* commands with open parts.

4. How to create a solid part from an open part.

Tutorial Instructions.

Start I-DEAS with a new model file.

Follow the instructions in this tutorial from the online Help Library.

Help Library, Tutorials,
Design, Surfacing,
Fundamentals,
Modeling Open Parts

At the end of this tutorial, you should be able to model a stove burner pan, which is most conveniently modeled using the open part technique.

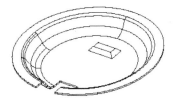

 Tutorial:
Performing Surfacing Operations

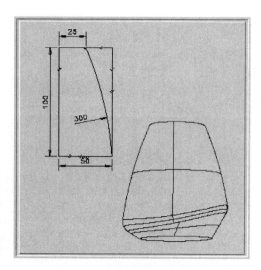

This tutorial demonstrates the *Variational Sweep* command. It also demonstrates how to use some of the direct surface creation and manipulation commands such as *Trim* and *Stitch*.

 Before you start:

1. This tutorial uses the *Variational Sweep* command, may not be available with all installations of I-DEAS.
2. You should know how to sketch and constrain wireframe curves.
3. You should understand the difference between open and solid parts.

 What you should learn:

1. How to use *Variational Sweep*.
2. How to project curves to surfaces, trim surfaces.
3. How to create surfaces using *Surface By Boundary*.
4. How to stitch surfaces together.

 Tutorial Instructions.

Start I-DEAS with a new model file.

Follow the instructions in this tutorial found in the online Help Library.

Help Library, Tutorials
Design, Design Part Modeling,
Fundamentals,
Performing Surfacing Operations

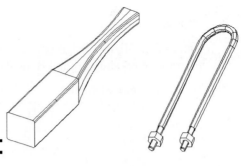

Workshop 5: Surfacing

In this workshop, you'll use some surfacing tools to model some proposed design alternatives for the handle of the pump designed in Chapter 2. That workshop did not show how the handle would be connected to the pivot block. One option would be to use two U-bolts. Here, you'll model a simple U-bolt as a swept part.

Another design alternative would be to extend the wooden pivot block as the handle, instead of bolting on the pipe. The wooden beam could be carved to produce an attractive handle, rather than just using a rectangular beam. This new pump handle will be designed using lofting.

Before you start:

1. You should already know the basic I-DEAS program mechanics such as getting into I-DEAS, changing tasks, and selecting icon commands.

2. You should know how to create solid parts from extrusions and primitive shapes.

3. You should know how to orient parts and the workplane.

4. You should know how to put away parts, and get them from the bin.

5. No previously existing parts are required for this workshop.

After you're done, you should be able to:

1. Create swept and lofted parts.

Workshop Instructions:

Enter the I-DEAS Design application, Modeler task. Either use the model file from Workshop 2A and B, or start a new one.

```
$ ideas <RETURN>
```

One way to fasten the handle to the Pivot Block in the pump designed in Workshop 2 is to use two U-bolts. We will use the Sweep method to create this part by sweeping a circular cross-section along a U-shaped path.

Start by sketching the U-shaped path on the workplane. Create two parallel vertical lines, and then fillet the top. (If the Navigator option *Linear Dimension* is on, a dimension will be created. If not, create a dimension between the lines.)

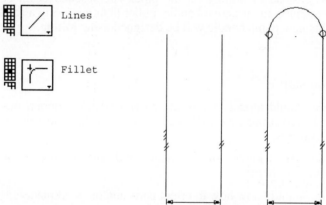

Lines

Fillet

Dimension the two vertical lengths, and match one to the other. Modify dimension values to be 60 mm horizontal, and 200 vertical.

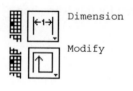

Dimension

Modify

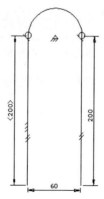

Create a circle anywhere on the workplane and modify its radius to be 5 mm. (Or use ▣ Options to set the radius.) This circle will be swept along the path to create the U-bolt.

Center Edge

Modify

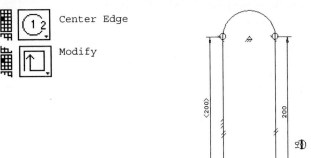

Sweep the circular section along the path.

Isometric View

Sweep

Put Away this part, naming it "U Bolt."

Put Away

U Bolt

✉ What about the threads?

It is usually not advisable to explicitly model threads on bolts, unless the purpose of the model is to study the form of the thread shape (such as to perform a finite element analysis of the stresses in the threads). What is often done is to show the inner or outer diameter of the thread so that the change in diameter will show up on a drawing. For example, here is one possible representation of the threads, without making a complicated model of all of the thread surfaces.

Another design alternative to explore is to extend the wooden Pivot Block to make the handle instead of bolting on a metal pipe. To shape the wooden beam into an attractive shape, we will draw several cross sections, and loft between them to create the new handle.

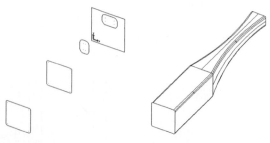

View from the front view, and draw a rectangle on the workplane about 150 mm square. (Watch the odometer for the values of W and H.)

Front View

Rectangle By 2 Corners

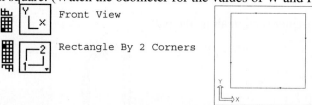

Copy this rectangle 400 mm in the -Z direction.

Move

(Pick one line) ▢▢▪ All

▢▪▢
Copy Switch **ON**
Translation **0 0 -400**
Number of copies **1**

Isometric View

Zoom All

Move the Workplane -700 mm in the Z direction.

Move

Translation **0 0 -700**

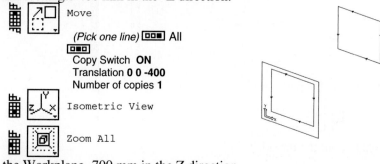

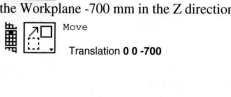

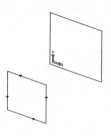

View from the front and draw a smaller rectangle as shown. *(Tip: To get this rectangle aligned with the first, which is in a different plane, you may use the* *Focus option to project a point to the workplane.)*

Front View

Rectangle By 2 Corners

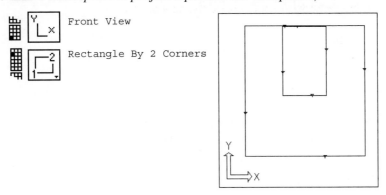

Move the workplane -300 mm in the Z direction.

Isometric View

Move

Translation **0 0 -300**

Zoom All

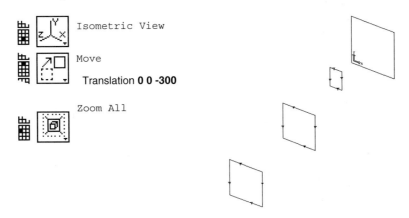

Draw another rectangle as shown.

Front View

Rectangle By 2 Corners

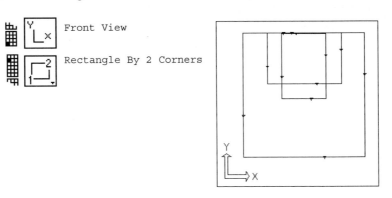

You should now have four rectangles spaced out as shown below in an isometric view.

Isometric View

Zoom All

Fillet each of the corners using a value of 10 mm on each corner. Hint: pre-select all corners before selecting the Fillet command.

Pre-select one corner, then use ▭ *All*

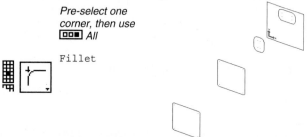

Fillet

Use the *Loft* command to create a new handle from these sections. Be careful to pick each rectangle in the same location. (The arrows should all be pointing in the same direction, clockwise or counterclockwise.) You will need to use F1 and F2 to zoom in on each curve to pick them. After picking all four, press Return. Modify the loft definition to break the surface at the second section.

Loft

Pick cross section curve
▭ pick 1st section ▭ Done
▭ pick 2nd section ▭ Done
▭ pick 3rd section ▭ Done
▭ pick 4th section ▭ Done
▭ (Done)
Options...
  Tangency
Pick section at which to modify tangency
  (Pick second section.)
Tangency **Non Tangent (Break)**
  👁👁 (Preview)
OK
▭ (Done)
[OK]
[OK]

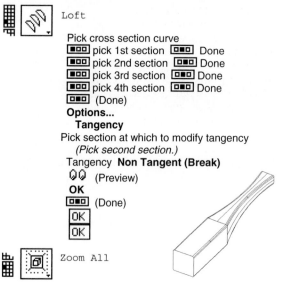

Zoom All

Save your model file and exit when you are finished. You may want to save your parts in a library or a file.

Chapter 6
Assemblies and Mechanisms

Assembly

```
Design Application Tasks
    Modeler
    Assembly
    Drafting
    Mechanism Design
    Harness Design
```

The Assembly task of the Design application is used to build assemblies out of the parts created in the Modeler task. Assemblies are used to visualize the complete assembly of parts, to calculate interferences, to calculate mass properties, and to animate configurations of the assembly.

In the Assembly task, the same part can be instanced many times in the assembly without creating copies in the database. Also, if the assembly of parts has multiple configurations (orientations of parts), the configurations can be efficiently stored without storing any extra part geometry.

As you begin a new design, the relationships and functions of the parts in the assembly are often defined before the details of each part are modeled. In this case, you may want to create very simple part models to put into an assembly. Later, as each part is refined, the updated parts can replace the simple parts in the assembly.

To use the Assembly task effectively, it is important to understand the concepts of assembly hierarchy, instances, and sub-assemblies.

Instances

When you place a part in an assembly, this is referred to as an "instance" of the part. If another copy of the same part is placed in the assembly again, the software does not make another copy of the geometry. Instead, another instance of the same part is created. Each instance of a part gets its geometry from the part stored in the bin, but contains new information about the orientation of the part. This makes the assembly process very efficient in terms of disk space, because the part geometry is only stored once, no matter how many times the part is used in the assembly. If the part is changed in the Modeler task, all instances of that part will change.

Assembly Hierarchy

The hierarchy of an assembly can be pictured as an inverted tree. The parent assembly contains instances of parts of assemblies. Each of the assemblies in this tree may also contain instances of parts or other assemblies, and so on, until the lowest level of each branch of the tree is reached. This hierarchy can have as many levels as needed.

After you have created an assembly of parts, it can be used as a sub-assembly in yet another assembly. The instance concept is also used here. If a sub-assembly is used multiple times, multiple instances "point" to the stored sub-assembly, but do not make duplicate copies in the database. An assembly, then, is a collection of instances of parts and/or assemblies.

In the workshop at the end of this chapter, two instances of the Offset Link will be made into an assembly called the "Link Assembly." This assembly will be used as one of a number of sub-assemblies in the overall assembly model. It is convenient to build the assembly this way, since the link assembly always moves as a unit. Defining the assembly using this sub-assembly will make it easier to define the orientations of the pieces later. The complete assembly hierarchy for the pump model is shown below.

```
┌─────────────────────────────────────────────────────┐
│              Pump Assembly Hierarchy                  │
│                                                        │
│   Pump Assembly                                        │
│       Link Assembly                                    │
│           Offset Link (part)                           │
│           Offset Link (another instance of same part)  │
│       Piping Assembly                                  │
│           Well Casing                                  │
│           Well Seal                                    │
│           Tee                                          │
│           Guide Pipe                                   │
│           Output pipe (optional)                       │
│           Elbow (optional)                             │
│           Base (optional)                              │
│           Pad (optional)                               │
│       Handle Assembly                                  │
│           Handle (part)                                │
│           Pivot Block (part)                           │
│       Pivot Support                                    │
│       Stroke Slide                                     │
└─────────────────────────────────────────────────────┘
```

The assembly hierarchy is displayed and modified with the command *Hierarchy*. This command displays the hierarchy in a form that lets you build the assembly from the top down or from the bottom up. When you start "with the paper blank," you will first start by adding a parent assembly. This forces you to give the assembly a name. At this point, this assembly (even though nothing is in it yet) will show up in the bin, if you use the command *Manage Bins*.

 Hierarchy

You can build upwards with the button *Add Parent* or build downwards with the buttons *Add To*, *Add Empty Assembly* and *Add Empty Part*. *Add To* will let you get parts or assemblies from a bin, library, or catalog.

Managing Assemblies

Assemblies are stored in the same bins as parts, as covered in Chapter 3. One column in the *Manage Bins* form lists the type of entity, showing whether it is a part or an assembly.

Assembly Constraints

The *Constrain Instances* icon displays a subpanel of icons which allow you to create permanent relationships between instances. Using constraints, rather than just locating instances in space with *Translate* or *Rotate* commands creates a permanent relationship between instances. If the position of one instance is changed or its dimensions change, the position of mating instances will automatically be updated.

 Constrain
& Dimension

I-DEAS uses a system of VGX constraints which use the same commands used for 2D constraints such as *Parallel*, *Perpendicular*, *Tangent*, *Dimension*, and *Coincident & Collinear*. Older versions of I-DEAS used relationships such as Face_To_Face and Line_To_Line constraints. The advantage of the VGX technology is that rather than following a sequence of prompts in one command to constrain all six DOF between two instances, each DOF can be constrained in a more intuitive manner. This also allows more flexibility in how instances are constrained to more than one instance.

You will normally also want to ground instances that should *not* move. Otherwise, you won't have control over which instance the program will move when part dimensions are changed. Use the command *Lock*, either to lock an instance to another instance, or to lock an instance relative to its parent assembly. A related command is *Fuse*, used to fuse all instances in a subassembly into one rigid body.

To help understand the constraints in an assembly, the *Browse Relations* command gives a graphical picture of the constraint network, something like the history tree. The circles represent instances, the curves between them represent each constraint. Selecting a constraint on the form will highlight the corresponding entities in the Graphics window.

 Browse Relations...

To further understand relationships, dimensions can be animated. (*Animate Dimensions* replaces the *Nudge* command in older versions of the software.) Another useful command is *Show Free*, to help find unconstrained degrees of freedom between instances.

Configurations

Different configurations of the assembly can be created to display animations of part motions or to check interferences of parts in different positions. A configuration contains the orientations of each instance in the assembly. The icon command to create configurations and to select which configuration to use is *Manage Configurations*.

 Manage Configurations

When you first create an assembly, it will have one configuration, named Config1. To create new configurations, use the *New* button on the *Manage Configuration*s form. Select the new configuration as the one you want to use. Then use the *Move* and *Translate* commands or modify constraint values to orient the instances where they belong in this new configuration.

When an instance's orientation (translation or rotation) is changed, only that particular instance of that part changes. This is different than translating the part itself. If you move or rotate the part in the Modeler task, all unconstrained instances will move or rotate.

A part taken from a library as a referenced part may not be modified in any way, including changing the part's location. However, instances of the part in the assembly can be moved.

Sequences

A "sequence" is a list of configurations. This sequence can be animated. Using the Mechanism Design task, it will be possible to have this sequence automatically created for steps of motion of a kinematic mechanism. This will be covered later in this chapter.

 Create Sequence

Animation

There are two types of animation. Normal animation of a sequence lets the configurations of parts change between frames of the animation. A more general animation allows you to change views and display options between each frame, as well as configurations. To use this more general animation, you need to store tables of view parameters for each frame of animation with the *Manage Views* command. This allows the type of animation where you can "fly over" the assembly. When you select the *Animation* command, the default is to pick a pre-defined sequence to animate.

Listing Information for an Assembly

Several types of analysis can be performed on assemblies. These include mass properties calculation, interference checking, and measuring distances and coordinate locations. Interference checks list information about which of the instances interfere, just touch, or do not interfere.

Physical Properties

The physical properties calculated for an assembly using the *Properties* command will reflect the physical properties of each part, using the current assembly orientation. You can assign a material to each part in the assembly in the Modeler task, so that the correct mass properties of each assembly are calculated.

There are some times when you may know the mass properties of a part or sub-assembly more accurately than the geometry is modeled. (You may have weighed a sub-assembly on a scale.) In this case, you can override the mass properties for a part or assembly by using the *Properties* command to enter rather than calculate mass properties.

When you compute assembly mass properties, it is important to check the calculation of mass properties for each instance of the assembly. A common mistake is not assigning a material or entering a valid density for each part in the assembly. This can be checked by generating a report using the *Properties* command.

Assembly Construction Operations

Construction operations on assemblies allow you to cut instances with other instances or with planes. These can be useful for making cut-away drawings, for using instances as tools to cut other parts in the assembly, or to create profiles for use with tolerance analysis.

Note that construction operations change the assembly definition. New entries to the bin will be added if parts change.

Assembly Features

Assembly features are modifications to parts made at the assembly level. This might be used, for example, when parts are otherwise the same, but are modified for assembly, such as drilling holes for final assembly.

For instruction on creating assemblies, see the tutorials and workshop at the end of this chapter.

Mechanism Design

A mechanism is an assembly of rigid bodies with specific degrees of freedom constrained relative to ground, or to other rigid bodies.

In I-DEAS 8 and above, there is both a Mechanism Design task in the Design application and a Mechanism Simulation task in the Simulation application. In the Mechanism Design task of the Design application, you can define a kinematic mechanism and calculate its motions as a function of time. In the Simulation application, you can also simulate mechanisms driven by forces and calculate output forces. This Student Guide will primarily focus on Mechanism Design.

The assembly of parts is first defined in the Assembly task. The Mechanism Design task adds joints to the system, defines inputs to the mechanism, and solves for the kinematic motions.

When the mechanism is solved, configurations will automatically be stored for the specified steps of the motion. These configurations can be used to perform any of the assembly checks described earlier in this chapter, including interference checks. You can also graph motion in an XY format as a function of time.

Mechanisms can be defined and solved within the Mechanism Design task or solved with an external solver such as ADAMS™.

Each link in the mechanism is called a "rigid body." Each rigid body has six degrees of freedom, with three translations and three rotations possible. A rigid body can be any instance within the assembly. If a sub-assembly is selected as a rigid body, all children of that sub-assembly will move as a unit. The internal solver requires that one rigid body be "grounded." This restriction may not apply to all external solvers, but many solvers will require this.

A "marker" (reference coordinate system) is placed on a rigid body to define the location and orientation of joints and loads. The definition of a marker contains its location, its orientation, and the rigid body to which it belongs. (A marker may also be referred to as a "reference triad.")

The rigid bodies are connected by "joints." Typical types of joints are pins (revolute joints) or sliders (translational joints) which constrain degrees of freedom between the connected rigid bodies.

Joints

A joint allows one or more degrees of freedom of motion while restraining others. Each freedom of motion has an associated "joint variable." If the value of every joint variable in a mechanism is defined, the position of every rigid body is defined. The joint variables can be plotted after solving for the motion of the mechanism.

A joint connects two reference triads on different rigid bodies. The hierarchy of a joint is shown in the following table.

Hierarchy of a Joint
Joint
Marker 1
Location
Orientation
Rigid Body
Instance (Part or Assembly)
Marker 2
Location
Orientation
Rigid Body
Instance (Part or Assembly)

Unless you are creating a mechanism that requires gears or complex constraints, you may never have to explicitly define rigid bodies and markers. When you ask I-DEAS to create a joint using the icon commands, it will automatically create the required marker and rigid body definitions for you.

Depending on the type of joint, certain restrictions will apply to the orientation of the coordinate systems of the two markers. A verify option is available to check for valid joints.

Joint types that can be defined include revolutes (pins), sliders (translational joints), ball joints, and cylinder joints. A revolute joint allows rotation about one axis. The joint variable for a revolute joint will be the angle of rotation (in degrees). The joint variable for a translational joint is motion in one translational degree of freedom. A ball joint has three joint variables, rotations about three axes. You must be careful to avoid configurations that allow the third joint variable to approach 0 or 180 degrees, which will cause the spin axes of joint variables 1 and 2 to become collinear, and cause numerical problems. A cylindrical joint has two joint variables, translation and rotation.

The Constraints icon shown below displays a panel of icons to create revolute, translational, cylindrical, spherical, universal, planar, cams, and other types of joints.

Constraints

Revolute Joint

Functions

Functions are used for entering motions as functions of time or as functions of joint motions to be used in a load case. A motion function at one joint is often all that is necessary to drive the mechanism.

Functions can be entered as evaluated expressions (e.g.: SIN(2*T)) or by keying in function values at the keyboard, using cursor entry from a grid, or by fitting a cubic spline to an existing function. After you enter a function, you can check it by plotting or listing its values. When a function expression is plotted, don't worry that by default the function is evaluated over a larger or smaller time range than you will use when you solve the mechanism. The ending time for the mechanism solve does not have to be the same as the range of the function expressions.

A function defining motion input should start at the same value as the joint variable of the starting configuration of the joint where the motion input will be applied.

The abscissa (X axis) of a function can be time or a joint variable. To create a function to drive the motion of the mechanism at a joint, create a function with time as its abscissa. A function defined using a function expression can contain more than one joint variable in the expression.

Functions can also be grouped for your convenience. By default, all functions you create are kept in a group called "User Defined." Functions can also be plotted and manipulated with math operations such as integration, differentiation, addition, and subtraction.

 Create
Functions

The Functions icon brings you to the Functions menu to create and modify functions. (Make sure menus are turned on, with the option "All" selected.)

Motions

 Create
Motion

Motions are created using the functions created as described above. Creating motions and forces applies these functions to specific joint variables.

Ground Instances

At least one instance (one part or a sub-assembly) must be grounded before the solution with the *Ground Instance* icon.

 Ground
Instance

Mechanism Solution

 Solve

When you request a mechanism solution using the Solutions icon, first select the solution program, the I-DEAS internal solver, or some other external solver.

The Load Type is also selected from this form. It is set to Motion input by default. The motions selected become the Load Case that will be used for the Solve. This set of motion inputs can be stored with the button Manage Load cases.

The *Output Selection* button sets flags to control the types of mechanism results desired. By default, all results will be stored during the solution. These results include body motion functions.

The button *Verify Mechanism* will check the joints in the mechanism. The output in the List window will tell you the type of each joint and the instances connected.

Display the Results

After the solution is complete, the results specified will be stored as response functions that you can plot, and as a sequence of configurations that you can animate or use to perform other assembly checks such as interference checking.

 Be careful however, if you have created assembly constraints. Although the mechanism solver will ignore assembly constraints, they are still active in the assembly. Turn off the Assembly Update toggle on the Update Options form before you select a configuration, or assembly constraints will immediately be solved, which will permanently alter the configuration as computed by the mechanism solver.

After the solution, you can also calculate other motions between different points using the Solution, Motion Analysis icon. This allows you to calculate positions, velocities, and accelerations between selected reference triads.

Steps in Solving a Mechanism

The following chart summarizes the basic steps in defining and solving a mechanism:

Steps For Mechanism Design
Modeler Task   Create parts
Assembly Task   Assemble system hierarchy

Mechanism Design Task
    Define Joints
    Create User-Defined Functions
    Create Motions using these Functions
    Ground (constrain) at least one instance
 	
    Solve
        Create Load Case with the desired inputs.
 	
    Plot Response Functions

Mechanism Design or Assembly Task
    Animate Configurations
    Interference Checks

Mechanism Simulation

 The Mechanism Simulation task has been part of the Simulation application since I-DEAS 8. This task has all of the capabilities described for Mechanism Design, plus simulating mechanism forces and dynamics. Force results can also be used as input to finite element models to calculate internal stresses.

 Create Force

Mechanism Simulation includes applied force functions and calculates force reactions at joints.

Dynamic Solves

Both Mechanism Design and Mechanism Simulation include an internal solver that lets you perform kinematic analysis. A kinematic analysis simulates the motion of a mechanical system as a function of time and determines the range of values for displacement, velocity, and acceleration of rigid bodies. In Mechanism Simulation, it also determines the reaction forces on constraints. However, the kinematics solver does not solve for mechanisms that have any extra degrees of freedom, and won't generate motions that result from initiating forces. Such mechanisms require a dynamic analysis. Both Mechanism Design and Mechanism Simulation can solve for dynamics problems. The software automatically determines if a dynamic solve instead of a kinematic solve is required. If there are extra degrees of freedom, it assumes that a dynamic solve is required.

Transferring Loads to Simulation

With Mechanism Simulation, you can also transfer results from a kinematic or dynamic solution to a finite element model for analysis of internal stresses caused by the mechanism motion and loading using the command *Load Transfer*. You can transfer loads as load sets at the peak (maximum) value or at specified time steps in the solution run.

Load Transfer creates nodal (point) forces at the joint locations, which don't always represent a realistic load application. If you're primarily interested in the stresses at a joint, consider creating new surface- or edge-distributed loads based on the value of the nodal forces.

You can also create a function ADF to describe the loads over the entire duration of the mechanism solution. This ADF function may be used in the Simulation applications in the Durability or Response Analysis tasks, or in the Test application.

Where To Go For More Information

For more information on these topics, see the following in the Help Library.

> *Help, Help Library, Bookshelf*
> > *Design User's Guide*
> > > *Assemblies – Creating and Managing*
> > > *Assemblies – Designing Within*
> > > *Assemblies – Working as a Team*
> > > *Mechanism Design/Mechanism Simulation Design Guide*

Also, use *Help, On Context* to directly look up any command in the Command Descriptions.

⟨ PARTING_SUGGESTIONS

Assembly Task

-Learn to use multiple instances and sub-assemblies properly to simplify changes to the assembly and reduce storage.

-Don't be confused between joints and assembly constraints. Assembly constraints, created in the *Constraint* sub-panel, are relations between parts in an assembly used by the Assembly task. These relations act like joints within the Assembly task, but do not define a joint to be used by the Mechanism Design task.

-When computing mass properties, be aware of the material properties used by each part. To change a part, you must check it out of the library (not reference).

-In I-DEAS Master Series, use libraries to share work with others. A whole assembly of parts can be checked into a library, or just individual parts.

-When you build an assembly from library entities, be aware if instances in the assembly are reference specific (RFS) or reference latest (RFL). By default, only the first level down in an assembly is reference latest. In some assemblies, such as the 2002 model of a car, the assembly should have the specific 2002 versions of the body parts, not the latest versions.

-Each assembly or sub-assembly (which is really the same thing) should have an assigned owner to maintain that assembly level. This person will decide when that assembly should be updated or not. The assembly will only change when the owner OK's the change by checking another version of the assembly back into the library. Assemblies should be updated from the bottom up.

☼ PARTING_SUGGESTIONS

Mechanism Design Task

-Choose the motion input (primary motion) and initial configuration to avoid indeterminate positions.

-Make sure you have a mechanism, not a structure. Creating a structure rather than a mechanism may be caused by misalignment of joint motions.

-The number of motion inputs must be equal to the number of possible motions.

-Carefully consider the selection of joint types, orientations, and degrees of freedom in the problem.

-Do not use redundant joints. For example, a yoke mechanism may physically look like two pins, but mathematically it is only one. A similar problem is caused by hydraulic actuators or shock absorbers. Due to the physical design, these devices are often kinematically redundant.

-Solver errors can be misleading. If you get an error, check any of the above conditions.

-Use the *Info* command to verify that you have entered joints and other mechanism data correctly.

-If you have created any assembly constraints, turn off the *Assembly Update* toggle on the *Update Options* form before selecting a configuration created by the mechanism solver.

 Tutorial:
Creating
Assemblies

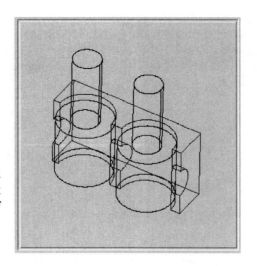

This tutorial is the start of a series of assembly tutorials. It teaches the basic concepts of assemblies.

Before you start:

1. You should be able to create parts.

2. You should know how to manage parts in the model file.

What you should learn:

1. How to create an assembly hierarchy.

2. The differences between assembly instances and parts.

3. How to move, constrain, and duplicate instances.

Tutorial Instructions.

Start I-DEAS with a new model file. This file will be used for the sequence of tutorials to follow.

Follow the instructions in the tutorials below.

Help Library, Tutorials
Design, Assemblies,
Fundamentals,
Creating Assemblies

 # Tutorial: Using Assembly Configurations

This tutorial continues with the last tutorial, to demonstrate the use of configurations.

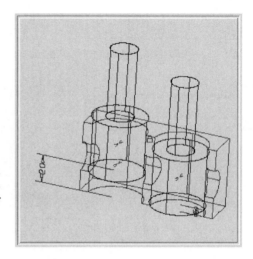

 Before you start:

1. You must have created the assembly in the previous tutorial.

2. You should understand the assembly hierarchy and instances.

 What you should learn:

1. How to find and remove assembly constraints.

2. How to create assembly configurations.

3. How to use configurations in a multi-level hierarchy.

 Tutorial Instructions.

Either start with a copy of the model file from the last tutorial, or start with a new file and get the assembly from a library, as instructed in the tutorial below.

Help Library, Tutorials
Design, Assemblies,
Fundamentals,
Using Assembly Configurations.

The modified assembly from this tutorial is not required for following tutorials, but keep the original assembly from the previous tutorial.

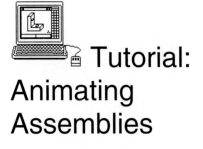

Tutorial: Animating Assemblies

A sequence of configurations can be animated to show motion.

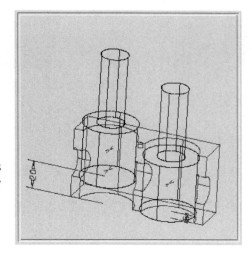

Before you start:

1. You must have the parts and assembly from the first assembly tutorial.

2. You should know how to create assemblies and configurations.

What you should learn:

1. How to create assembly sequences.

2. How to animate assembly sequences.

Tutorial Instructions.

Either start with a copy of the model file from the first tutorial in this chapter, or start with a new file and get the assembly from a library, as instructed.

Follow the instructions in the tutorial below.

Help Library, Tutorials
Design, Assemblies,
Fundamentals,
Animating Assemblies

Save the original assembly for the next tutorials. The modifications to the assembly from this tutorial are not required for following tutorials.

Tutorial: Modifying Assemblies

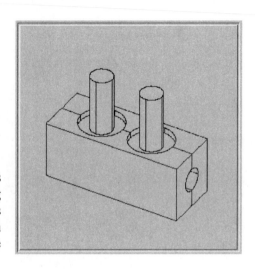

This tutorial demonstrates some techniques of modifying parts in an assembly, such as making a reflected part which is an associative copy of the original.

Before you start:

1. You must have the parts and assembly from the first assembly tutorial.
2. You should know how to create assemblies and configurations.

What you should learn:

1. How to check assembly interferences.
2. How to modify parts in an assembly.
3. How to create an associative copy.
4. How to use the *Browse Relations* command to display relationships between instances.

Tutorial Instructions.

Either start with a copy of the model file from the first tutorial in this chapter, or start with a new file and get the assembly from a library, as instructed in the tutorial below.

> *Design, Assemblies,*
> *Fundamentals,*
> *Modifying Assemblies*

Save the first model file for the next tutorials. The modifications to the assembly from this tutorial are not required for following tutorials.

For more details on associative copy, see the tutorial:

> *Design, Assemblies,*
> *Advanced Projects,*
> *Using Associative Copy* or
> *Modeling a Mold for a Plastic Part*
> (Student Edition 1.0 or prior to Master Series 7)

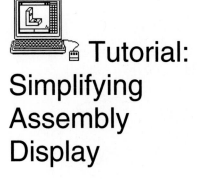

Tutorial: Simplifying Assembly Display

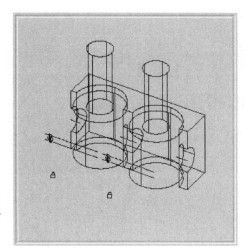

This tutorial demonstrates how to modify the visibility of assembly instances.

Before you start:

1. You must have the parts and assembly from the first tutorial.

2. You should know how to create assemblies and configurations.

What you should learn:

1. How to hide and show assembly instances.

2. How to suppress assembly instances.

3. How to abstract and color-code assembly instances.

4. How to prune assembly instances. (If you have access to libraries.)

Tutorial Instructions.

Either start with a copy of the model file from the first tutorial in this chapter, or start with a new file and get the assembly from a library, as instructed in the tutorial below.

Help Library, Tutorials
Design, Assemblies,
Fundamentals,
Simplifying Assembly Display

Save the first model file for the next tutorials. The modifications to the assembly from this tutorial are not required for following tutorials.

 # Tutorial: Documenting Assemblies

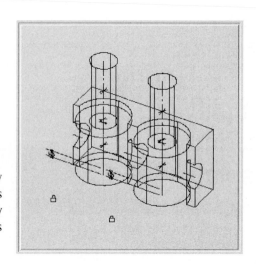

This tutorial demonstrates how to create a Bill of Materials (BOM) report for an assembly and how to calculate mass properties.

Before you start:

1. You must have the parts and assembly from the first tutorial.

2. You should know how to create assemblies.

What you should learn:

1. How to create an assembly Bill of Materials.

2. How to calculate assembly mass properties.

Tutorial Instructions.

Either start with a copy of the model file from the first tutorial in this chapter, or start with a new file and get the assembly from a library, as instructed in the tutorial below.

> Help Library, Tutorials
> Design, Assemblies,
> Fundamentals,
> Documenting Assemblies

Save the first model file for the next tutorials. The modifications to the assembly from this tutorial are not required for the next tutorials.

You may also want to try the related tutorial in the next chapter, to create an assembly drawing of this assembly.

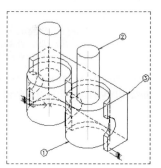

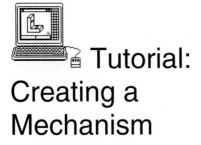

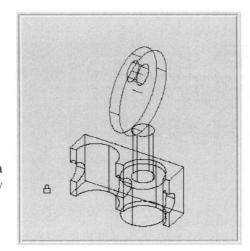

Tutorial: Creating a Mechanism

In this tutorial, you will add a cam to the previous assembly and create a mechanism.

Before you start:

1. You must have a version of I-DEAS with a license for Mechanism Design.
    (Mechanism Design is not included in the Student Edition 1.0)

2. You must have the parts and assembly from the first assembly tutorial.

3. You should know how to create assemblies and configurations.

What you should learn:

1. How to create a mechanism consisting of rigid bodies and joints.

2. How to apply motions to the mechanism and display the results.

Tutorial Instructions.

Either start with the model file from the first tutorial in this chapter, or start with a new file and get the assembly from a library, as instructed.

Follow the instructions in the tutorial below.

> *Help Library, Tutorials*
> *Design, Assemblies,*
> *Fundamentals,*
> *Creating a Mechanism*

This is the last of this series of tutorials. You may want to keep the original assembly, either to review one of these tutorials or to create an assembly drawing in a tutorial in the next chapter.

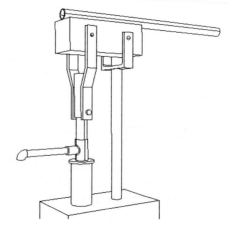

Workshop 6A: Pump Assembly

In this workshop, you will build an assembly out of the pump parts created in Chapter 2. In that workshop, you displayed the parts together on the workbench, but they were not related to each other as a system. That was an inefficient way to build the model, since you needed to create two copies of the Offset Link. In the Assembly task, if a part is used more than once, multiple instances of it can be used without duplicating the geometry in the model file.

The Assembly task will also let us animate different configurations of the pump. Animations and configurations will be expanded on in the next workshop, where we will let the mechanism solver calculate all the configurations of the assembly.

Before you start:

1. You should know how to create parts, manage their storage, and how to display them.

2. You must have the pump parts from Workshop 2 either in a library or a model file before you start this workshop.

After you're done, you should be able to:

1. Create an assembly of parts.

2. Create and store different configurations of an assembly.

3. Animate a sequence of configurations.

In the next steps, you will build an assembly of the pump parts as shown below.

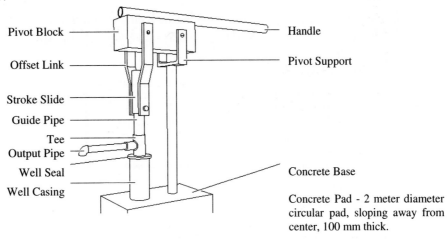

Pivot Block

Offset Link

Stroke Slide

Guide Pipe

Tee
Output Pipe

Well Seal

Well Casing

Handle

Pivot Support

Concrete Base

Concrete Pad - 2 meter diameter circular pad, sloping away from center, 100 mm thick.

You will create an assembly with a hierarchy as shown below. For some parts, such as the Pivot Block and Handle, it makes sense to group these together as a sub-assembly, because when they move, they move as a unit. Placing them in a sub-assembly will make it easier to position the parts in the assembly.

Pump Assembly Hierarchy
Pump Assembly
    Link Assembly
        Offset Link
        Offset Link
    Piping Assembly
        Well Casing
        Well Seal
        Tee
        Guide Pipe
        Output pipe *(optional)*
        Elbow *(optional)*
        Base *(optional)*
        Pad *(optional)*
    Handle Assembly
        Handle
        Pivot Block
    Pivot Support
    Stroke Slide

Each line in this assembly hierarchy represents an assembly (it makes sense to refer to these as "sub-assemblies"), or an instance of a part. The parts labeled "optional" were created as "On Your Own" exercises, and you may not have them in your model file. Refer to the picture on the next page to see how these parts are related.

Workshop Instructions:

Start the Design application, Modeler task either with a new model file if you have your parts in a library, or use an existing model file containing the pump parts.

```
$ ideas <RETURN>
```

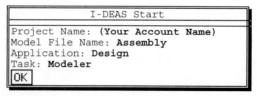

If you have a library of parts, get a reference of each of the parts in the Pump Parts library into your model file. (Don't include the Offset Link Copy.)

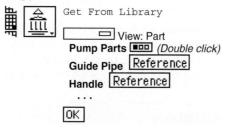

Put away any parts on your workbench, and switch to the Assembly task.

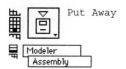

If you did not start in a new model file, make sure your workplane is aligned to the global coordinates. Use the command:

Bring up the Hierarchy form, and add a parent assembly named "Link Assembly."

 Hierarchy

 (Add Parent)
Link Assembly

Add an instance of the Offset Link to this assembly.

Hierarchy

Link Assembly ▣□□ *(Select Line)*

(Add To)
□□□ Get From Bin/Library
Offset Link
OK
Dismiss

Isometric View

Zoom All

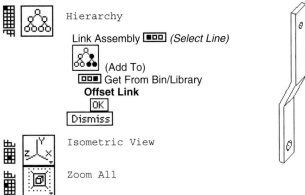

Rotate this instance of the Offset Link 180 degrees about the global Y axis. (Because this is an instance of the part, the part orientation as stored in the bin will not change, only the location of this instance will move.)

Rotate

Pick Entity to rotate ▣□□ *(Pick Offset Link.)*
Pick Entity to rotate (Done) □▣□
Pick pivot point (origin) □▣□
About Y
Enter Angle 180

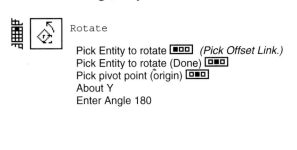

Add another instance of the Offset Link to the assembly.

 Hierarchy

Link Assembly *(Select Line)*

(Add To)

Get From Bin/Library

Offset Link

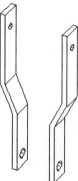

Put the Link Assembly inside a higher level "Pump" Assembly.

 Hierarchy

Link Assembly *(Select Line)*

(Add Parent)

Pump

```
Pump
    Link Assembly
        Offset Link
        Offset Link
```

Add (empty) sub-assemblies to the pump assembly as shown below.

 Hierarchy

Pump *(Select Line)*

(Add Empty Assembly)

Handle Assembly

(Add Empty Assembly)

Piping Assembly

```
Pump
    Piping Assembly
    Handle Assembly
    Link Assembly
        Offset Link
        Offset Link
```

Add the parts "Pivot Support" and "Stroke Slide" to the top level assembly. *Hint- use Control-Click to select multiple items.*

 Hierarchy

Pump  *(Select Line)*

(Add To)
Get From Bin/Library
Pivot Support
Stroke Slide

 Zoom all

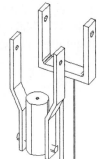

Add all the piping parts to the Piping Assembly.

 Hierarchy

Piping Assembly *(Select Line)*

(Add To)
Get From Bin/Library
Well Casing
Well Seal
Tee
Guide Pipe
Output Pipe*
Elbow *
* Optional, if you have these parts.

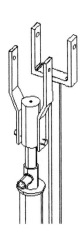

Add the parts to the Handle Assembly.

 Hierarchy

Handle Assembly *(Select Line)*

(Add To)
Get From Bin/Library
Pivot Block
Handle

Deselect All

 Zoom All

File
Save

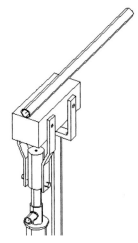

Now that the assembly hierarchy is complete, we will make and store some different configurations, showing the parts in different positions. Make a copy of the existing configuration, naming the new configuration "Angle 0". Make another copy named "Angle 10", and select this configuration as the active one to use.

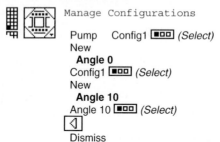

Rotate the Handle Assembly by -10 degrees about the X axis. (Use dynamic viewing to zoom in.)

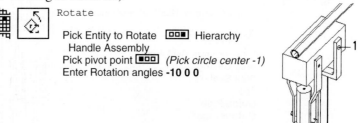

The Link Assembly rotates slightly as it translates up. To calculate how far it should rotate, we will measure the Z distance from the center of the upper hole in the Link to the mating hole in the Pivot Block. We will measure the Y length of the Link, between the lower and upper holes. The angle that the Link needs to rotate is A=ASIN(Z/Y).

Measure from the center of the upper hole in the Link to the center of the mating hole in the Pivot Block. Store the resulting Z value in a variable named Z. (The programmability commands used here are discussed more fully in Chapter 8.)

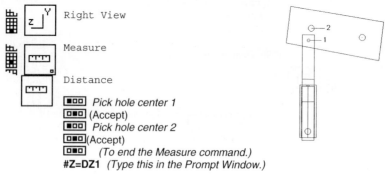

Measure the length of the Offset Link from the lower to the upper hole.

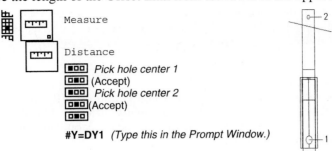

Measure

Distance

⬛⬜⬜ *Pick hole center 1*
⬜⬛⬜ (Accept)
⬛⬜⬜ *Pick hole center 2*
⬜⬛⬜(Accept)
⬜⬛⬜

#Y=DY1 *(Type this in the Prompt Window.)*

Calculate the angle. Output the value to the List window to check that you did it right. It should be a very small number, about -.5 degrees.

#ANGLE=ASIN(Z/Y)
#OUTPUT ANGLE

Rotate the Link Assembly by this angle about the lower hole center.

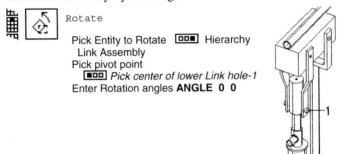

Rotate

Pick Entity to Rotate ⬜⬛⬜ Hierarchy
  Link Assembly
Pick pivot point
    ⬛⬜⬜ *Pick center of lower Link hole-1*
Enter Rotation angles **ANGLE 0 0**

Now we need to slide the Stroke Slide and the Link Assembly up to match the upper Link hole with the Pivot Block Hole. Measure the vertical distance between the mating holes in the Link and the Pivot Block. Translate the Stroke Slide and the Link Assembly by the Y distance between these holes.

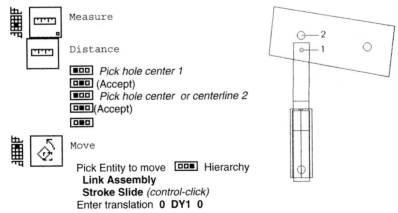

Measure

Distance

⬛⬜⬜ *Pick hole center 1*
⬜⬛⬜ (Accept)
⬛⬜⬜ *Pick hole center or centerline 2*
⬜⬛⬜(Accept)
⬜⬛⬜

Move

Pick Entity to move ⬜⬛⬜ Hierarchy
  Link Assembly
  Stroke Slide *(control-click)*
Enter translation **0 DY1 0**

Make the first configuration active, because you cannot select the active configuration into a sequence.

 Manage Configurations

Config1 ▣▢▢ *(Select)*

◁ Use

Dismiss

Create a sequence of the first two configurations. (If you are using I-DEAS 8 or higher, the steps will be slightly different. Use the *Manage Sequences* icon. Select the *Create Sequence* icon on the *Manage Sequences* form.)

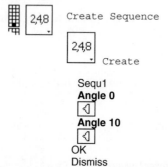 Create Sequence

2,4,8 Create

Sequ1

Angle 0

◁

Angle 10

◁

OK

Dismiss

Animate this sequence of configurations.

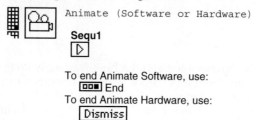 Animate (Software or Hardware)

Sequ1

▷

To end Animate Software, use:

▢▢■ End

To end Animate Hardware, use:

Dismiss

Save your model file.

 File

Save

In the next workshop, we will let the mechanism solver automatically compute all the configurations for a range of motion. Exit when you are finished or continue with the next workshop.

After learning about assembly constraints, you might also want to try adding constraints to the previous assembly.

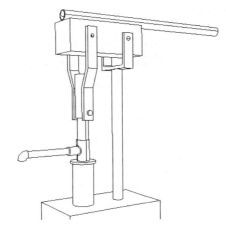

Workshop 6B: Mechanism

This workshop continues with the pump assembly created in the last workshop. The pump assembly will be defined as a kinematic mechanism. Joints will be added to define the connections between the instances of the assembly. The mechanism solver will calculate all the configurations of the assembly, so that this sequence of configurations can be animated.

Before you start:

1. You must have a version of I-DEAS with a license for Mechanism Design.

1. You should already understand assembly modeling concepts such as assembly hierarchy, instances, configurations, and sequences.

2. You must have the model file "Assembly" from Workshop 6A before you start this workshop. This model file must contain the assembly "Pump," created in that workshop.
   If you do not have this assembly, you must complete at least the first four pages of Workshop 6A to create the assembly hierarchy.

After you're done, you should be able to:

1. Add joints to an assembly and create a mechanism.

2. Solve the motion of the mechanism and graph the resulting functions.

Workshop Instructions:

Start I-DEAS with the model file used in the last workshop. Get into the Design application, the Assembly task.

```
$ ideas <RETURN>
```

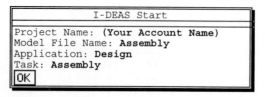

```
            I-DEAS Start
Project Name: (Your Account Name)
Model File Name: Assembly
Application: Design
Task: Assembly
OK
```

Check the hierarchy of the pump assembly. If it does not show up, you may need to Get the Pump Assembly.

 Hierarchy

Switch to the Mechanism Design task.

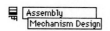
Assembly
Mechanism Design

Specify all the sub-assemblies to be rigid.

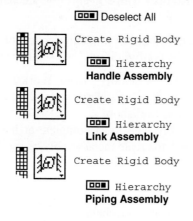

□□■ Deselect All

Create Rigid Body

□□■ Hierarchy
Handle Assembly

Create Rigid Body

□□■ Hierarchy
Link Assembly

Create Rigid Body

□□■ Hierarchy
Piping Assembly

Ground the Piping Assembly and the Pivot Support.

 Attach to Ground

(Pick well casing lower center point)

 Attach to Ground

(Pick pivot support lower center point)

Create a revolute joint between the Pivot Support and the Pivot Block. Pick the Pivot Support as instance 1 and the Pivot Block as Instance 2.

 Joints

 Revolute Joint

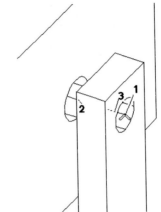

Pick joint origin *(Pick 1)*
Pick 2nd instance *(Pick 2)*
Pick Joint Z-axis
    (Pick 3, centerline)

(Accept the default direction of the axis. The way it points only causes a sign change on the direction of rotation.)

Create a similar revolute joint between the Offset Link and the Pivot Block.

 Joints

 Revolute Joint

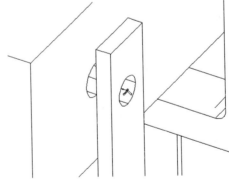

Create a revolute joint between the Stroke Slide and the Offset Link.

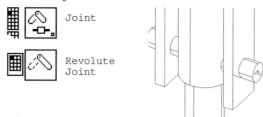

Joint

Revolute
Joint

Create a translational joint between the Stroke Slide and the Guide Pipe.

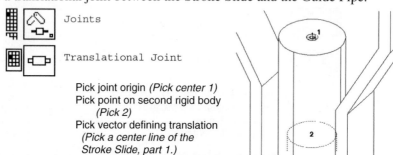

Joints

Translational Joint

Pick joint origin *(Pick center 1)*
Pick point on second rigid body
    (Pick 2)
Pick vector defining translation
    *(Pick a center line of the
    Stroke Slide, part 1.)*

Create a function for the pump driving angle vs. time. Create a sinusoidal
input motion with an amplitude of 20 degrees, and a period of 2 seconds.
(In 2 seconds, the sine function will complete one 360 degree cycle.)

Create Functions

Name: Function1
Add Function: CUBSPL
20*SIN(180*T)

 Plot Function

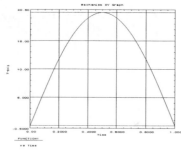

Create a Motion, applying the function just created to joint 1, the first revolute joint created above.

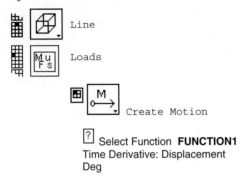

Line

Loads

Create Motion

? Select Function **FUNCTION1**
Time Derivative: Displacement
Deg

Solve the mechanism for 2 seconds of motion, with 24 steps.

 Solve

Steps: 24
End Time: 2
OK

To plot a result function, it will help to know which rigid body motion to plot. Use the *Mechanism Info* command to find out the name of the rigid body associated with the Stroke Slide. List information on joint 4, which will list the name of the attached instances. The name of the instances will include something like "RB5", meaning rigid body number 5.

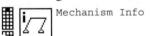 Mechanism Info

Pick Mechanism entity *(Pick the translational joint.)*

Display some of the output functions with the *Graph XY* icon.

Graph XY

Run1
RB5_Stroke Slide
Y Displacement

Add to plot

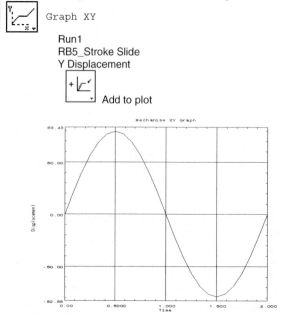

Animate the sequence created from Run1.

Animate Mechanism

Loop (on)

This is the end of this workshop. Feel free to try graphing other functions or animating using other display options.

The parts in the Pump Parts library will be used for following workshops, so don't delete them.

Save and Exit when you are finished.

Chapter 7
Annotation & Drafting

Documenting the Product Design

The complete documentation of a product includes more information than just the 3D geometry. Other information required to completely document a product design includes the following.

```
                    Product Documentation

    •Geometry (Topology)
    •Dimensions
    •Tolerances for each dimension
    •Geometric tolerances of features
        (Datums, GD&T feature control frames)
    •Manufacturing information
        (Surface finish, welding notation, etc.)
    •Inspection information
        (Key location points)
    •Assembly instructions
    •Product information
        (Materials, Suppliers, Part numbers,
        Bill of Materials, Revision history, etc.)
```

This chapter describes how this information can be documented using 2D and 3D methods.

2D vs. 3D Documentation

The product documentation listed above has traditionally been recorded on 2D paper drawings. The trend is toward both an electronic paperless environment and using a 3D representation rather than 2D drawings. The speed of this transition varies with different companies in different industries.

Companies tend toward either a "model-centric" or a "drawing-centric" philosophy of how their products are documented. The difference is not whether they use 3D models or 2D drawings or both. The question is which form contains the true released information. For example, a "drawing centric" company may perceive of 3D part models as a preliminary step to create 2D drawings, but the released 2D drawings are archived to document the final design. A "model-centric" company may hold a view that the 3D part model contains the master information, and the 2D drawings are only an intermediate form of communication to transmit information to suppliers who need printed drawings.

Which view a company holds will affect how they use 2D and 3D documentation. For example, manufacturing information such as GD&T feature control frames (which will be described later in this chapter) can be placed either on the 3D part model or added as annotation to the 2D drawing.

Although there are times when a 2D approach is necessary, there is a trend toward more "model-centric" companies, because any information attached to the 3D part model is available for multiple uses.

Individual Roles

Just as companies have a model-centric or a drawing-centric view, individuals within these companies have a different view, depending on their uses for product notation. This will lead to different practices in how additional manufacturing and product information is created, documented and used.

Individuals with an Industrial Design role tend to focus on the form and aesthetics of a design. They may work entirely with 3D models and not be interested in 2D drawings or manufacturing information. Individuals in this role are more interested in the nominal geometry than in tolerance deviations.

Those in an Engineering role focus on the function of the product. They may mainly use 3D models, but occasionally generate an informal 2D drawing to get a prototype made. When defining allowable tolerances, they will understand this information in a 3D context. In their view, because additional product information defines the function of a product, they will tend to want it available with the 3D models.

Manufacturing users tend to focus on manufacturing processes and assembly fit. They work with surface finish, dimensions, and tolerance. This information is traditionally presented in 2D drawing views. The "model view" feature in I-DEAS, discussed later in this chapter, will help these users make the transition to using 3D models to view information traditionally displayed in 2D.

Drafting users typically focus on the appearance of 2D documentation. They may create 2D drawings from existing parts and assemblies, or create drawings from scratch. In their view, dimensions and other notation belong to the drawing view instead of the 3D part.

Because users in each of these roles have different uses of product information, they may prefer to create it at different phases of the development process and store it in different ways. The software allows for different methods of creating and presenting this information to accommodate different needs.

Manufacturing Tolerance

One measure of quality in a manufactured product is the minimum amount of variation (tolerance) between each part produced. Manufacturing tolerance affects how parts will fit and function in an assembly. Proper tolerance also ensures that parts will behave as analyzed for stress and deflection.

Dimensional and Geometric Tolerance

The standard method of specifying tolerance is to place a +/- value on a dimension. There are times however, when this does not give enough information. For example, if you look at the dimensions of a simple block with a hole as shown below, what do the tolerances mean if the hole is not exactly perpendicular to the face? How round is the hole? The tolerances also do not state how accurately faces of the block must be machined parallel and perpendicular to each other, which also is a factor in controlling the hole feature.

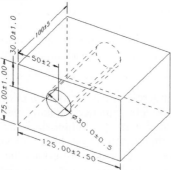

Another type of tolerance refers to the *form* of a part or its features. This type of tolerance cannot be stated as a variation of a dimension value. For example, if a part has a drilled hole, it is important to tell the manufacturing planner how accurately the hole feature must be machined in relation to other surfaces, and how round the hole must be. A standard way to convey this type of information is with GD&T (Geometric Dimensioning and Tolerancing) frames and datums.

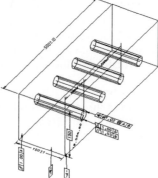

Before documenting tolerances, you need to determine what values to use.

Tolerance Analysis

Before tolerances can be documented, analysis should be done to define appropriate tolerance values. Controlling tolerance is more than doing a careful job on the factory floor or placing overly tight tolerances on dimensions. The most important place to control tolerance is in the design, to minimize the impact of uncontrollable tolerance.

The way a part is dimensioned defines how it should be manufactured. For example, consider a part with a row of holes as shown below. The tolerance of the last hole will be very different if holes are dimensioned to each other as shown on the left, rather than dimensioning each hole back to one datum, as shown on the right.

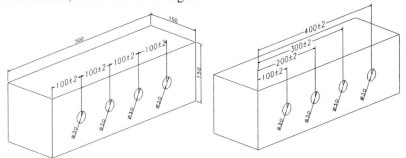

This illustrates that the way dimensions are applied is an important factor in tolerance stack-up. Some of the manufacturing planning has begun as soon as dimensions are applied to the model. The use of tolerance analysis can be a good teaching tool to illustrate the proper part dimensioning.

The *Tolerance Analysis* icon panel in the Modeler task lets you analyze tolerance stack-up on a two-dimensional model using planar sections of parts or assemblies. This tolerance analysis model is based on variational geometry previously described in Chapter 4. It allows you to predict worst case and statistical tolerances of a given reference dimension, taking into account all possible variations of other dimensions and constraints in the tolerance model.

To create a tolerance model, first create 2D wireframe sections to be used in the tolerance model. These sections can be sketched, extracted from a part, or created by cutting through a part or assembly. Geometric constraints and dimensions applied to this geometry become the "tolerance model."

Another type of dimension you need to create for tolerance analysis is a "reference" dimension. These are dimensions placed on the section for reference only; they do not drive the variational geometry. These dimensions can be picked as "dimensions to be analyzed" when you solve the tolerance model. To create a reference dimension, begin as if you were creating a normal dimensional constraint, but pick *Reference* out of the pop-up menu before you accept the location of the text for the dimension.

Each dimension and constraint in the tolerance model has some manufacturing tolerance within which it can be manufactured. Variation in each dimension can affect the reference dimension being analyzed. One computed result is the worst possible minimum and maximum "stack up" variation of the reference dimension being analyzed. Another result is the statistical upper and lower tolerances and the acceptance rate where the given reference dimension will fall inside the acceptable zone defined by the tolerances allowed on the reference dimension.

If the analysis shows that too many parts would fall outside the allowable range, the remedy could be to (1) tighten the tolerances on the individual dimensions in the model that contribute to the variation, (2) allow more variation in the reference dimension, or (3) change the geometry so that the reference dimension is less sensitive to individual tolerances that cannot be controlled as well. The sensitivity of each dimension in the model can be listed to help in deciding which tolerances should be adjusted.

For example, the following section was generated from a cross section of a part. The lower left point was grounded, and other constraints were placed on the profile to fully constrain it, in a manner consistent with the way it would be manufactured. After the section was fully constrained, a reference dimension was added across the top of the slot to analyze how the manufactured dimension is affected by other tolerances in the model. This reference dimension was picked as the dimension to be analyzed, and a standard tolerance analysis executed. With the tolerance variables given to each dimension, the analysis showed that the largest contributor to the final value was the tolerance angle allowed on the parallel constraint between the slot bottom and the base.

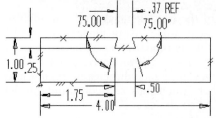

```
Tolerance Analysis Results:
Name: D15      Nominal: 0.3660255
Stat. Upper Tolerance: 0.06251235
Stat. Lower Tolerance:-0.06251235
Upper Tolerance:        +0.02
Lower Tolerance:        -0.02
Worst Up_Stk:  +0.11380357
Worst Low_Stk: -0.11380357
Acceptance Rate:       66.28609
```

The results of the tolerance analysis can help in applying meaningful dimension tolerances and geometric form tolerances to the model.

Applying Dimensional Tolerances

After you analyze tolerances to know what values are allowable for the function of a part or assembly, document the required tolerances using dimensional tolerances and form tolerances. As discussed earlier, both types of tolerance can be documented either on the 3D part model or on a 2D drawing, however the trend is toward using the 3D part to completely document the design.

In I-DEAS, you may add the tolerance to any linear or angular dimension with the *Modify* command. Click on the icon on the form to expand the form to show the fields for tolerance definition.

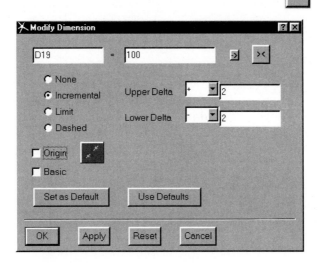

Product and Manufacturing Information (PMI)

In I-DEAS, several types of Product and Manufacturing Information (PMI) can be added to parts and assemblies. These include standard Geometric Dimensioning and Tolerancing (GD&T) feature control frames, datum targets, measurement points, locator symbols, surface finish, welding symbols, notes, and annotation dimensions.

To create each type of annotation, you pick the part, pick the associated surfaces or edges, locate the attachment point, and specify the plane of the text and the direction of the text.

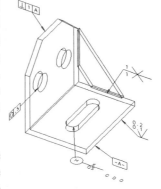

Document and Annotation Dimensions

The command *Modify, Show Dimensions* will display all the key dimensions used to create a part. These dimensions are displayed temporarily, and disappear when you use the *Update* command.

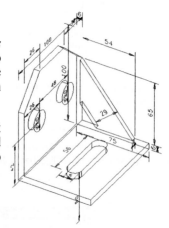

These key dimensions typically represent the engineering intent behind the part, and may be different than the way you want to document and manufacture the part.

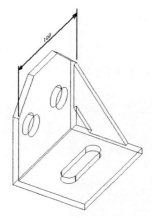

You can create annotation dimensions directly on the 3D part. These dimensions do not "drive" the part like key dimensions, but they will update when changes are made to the geometry of the part. By default, they are yellow so that you can tell the difference between them and key dimensions.

 A concept introduced in I-DEAS 8 is a "document" dimension. Dimensions flagged as document dimensions remain visible when the part is updated, To indicate a document dimensions turn on the *Document Dimension* toggle using the *Appearance* command.

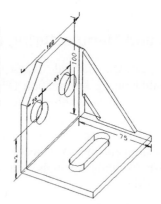

When a driving dimension is converted to a document dimension, it remains a key driving dimension. You can modify its value with the *Modify* command. When you create a drawing, document dimensions can be displayed independently of key dimensions.

Notes

Text notes can be applied to parts to communicate product information to users of the part. These notes can also contain a URL address which could link to a supplier of the part or other types of documentation.

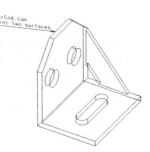

Weld Symbols

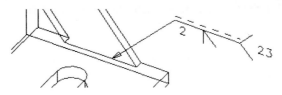

Weld symbols can also be applied directly to the 3D part.

The information above and below the line of the weld symbol defines the two sides of the weld. In the figure above, a fillet weld is specified on the arrow side of the weld.

Spot weld symbols are added by first creating reference points to define the location of the spot welds.

Spot weld symbols don't display as much information in the symbol itself. Use the *Info* command to list the content of the weld information.

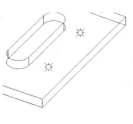

Surface Finish

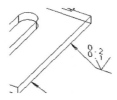

Surface finish symbols define required surface quality to satisfy design intent. Surface finish should only be specified for those surfaces where finish quality affects function.

GD&T Frames and Datums

Geometric Dimensioning and Tolerancing (GD&T) frames are a special "language" to describe manufacturing tolerances on part features that cannot be specified as just a +/- tolerance on a dimension.

Datum feature frames and datum targets indicate key reference surfaces and points. They are normally applied to surfaces used to mount the part for machining, to locate the part in an assembly, or to mount the part on a coordinate measurement machine (CMM) for inspection. For example, the symbol below placed on a drawing labels a surface as surface "A" that will be referred to in GD&T feature control frames. The form on the left is used in the ANSI standard, and the form on the right for the ISO standard.

$$\boxed{-A-} \quad \text{or} \quad \boxed{A}$$

Datum feature frames can be located on existing part geometry or reference geometry. Reference geometry can be planes that are attached to the part for reference and dimensioning, but are not a physical component of the part. Reference geometry may be particularly important to use when the part has no flat surfaces, or when the datum reference may lie off of the part.

Feature control frames identify geometric relationships between part surfaces. Most feature control frames refer back to one, two, or three of the datum references, such as to state the required tolerance on perpendicularity or parallelism to the datum surface. Some feature control frames do not reference other datums, but only describe attributes of a surface or feature, such as flatness.

GD&T feature control frames contain three parts. For example, to identify that a surface must be parallel within a tolerance to datum reference A, a feature control frame like the following would be used.

$$\boxed{// \mid 0.005 \mid A}$$

The first piece of information in a feature control symbol is always a geometric characteristic symbol. The chart on the next page summarizes these characteristic symbols.

The next block contains tolerance values and optional modifying symbols.

The third block contains datum references that this feature control frame refers to. This will depend on the geometric symbol in the first block. Some symbols do not use a datum reference and some always must use one. Others may optionally use a datum reference. This is summarized in the chart on the next page.

There will be slight differences in the construction of the feature control symbol and the datum symbol, depending on whether the ANSI or the ISO standard is being used. By default, the ISO standard will be applied. To change it, use the *Appearance* icon, and either pick a GD&T symbol to modify, or pick *Defaults*, then *Annotation*. The form that appears has menus to select the standard to use.

Geometric Characteristic Symbols

Symbol	Name	Datum	Description
∠	Angularity	Always	The amount that the surface is allowed to deviate from the specified angle.
○	Circularity	No	The amount that any point on a circle may deviate from a perfect circle.
◎	Concentricity	Always	The amount that the axis point on any cross section of a revolved surface is allowed to deviate from being concentric (same center) with the datum reference.
⌭	Cylindricity	No	The amount that any point on a cylindrical surface may deviate from a perfect cylinder.
▱	Flatness	No	The amount that any point on a surface may deviate from a perfectly flat plane.
//	Parallelism	Always	The amount which a surface may vary from parallel with the designated datum reference.
⊥	Perpendicularity	Always	The amount that a surface may deviate from being perfectly parallel to the designated datum reference.
⊕	Positional Tolerance	Always	The amount a feature may deviate from the stated dimension from a datum reference.
⌒	Profile of a Line	Optional	The amount of deviation of a profile shape from the specified shape.
⌓	Profile of a Surface	Optional	The amount of deviation of a surface from the specified shape.
↗	Circular Runout	Always	Runout is measured with a gauge as the part is rotated about the circle center. Circular runout is a measure of both the location and the roundness of a circle.
↗↗	Total Runout	Always	Total runout is a measure of the deviation of any point on a cylinder as the part is rotated about the cylinder axis.
—	Straightness	No	The amount a line on a surface may vary from a perfectly straight line.
≡	Symmetry	Always	The amount of deviation from perfect symmetry.

Modifying Symbols

In some cases, the following modifying symbols will be added to the tolerance field of the feature control frame.

The symbol for diameter $\varnothing$ always is placed in front of the value, if used.

The symbols Ⓜ, Ⓛ, and Ⓢ are used to state that a tolerance applies at the maximum material condition (MMC), the least material condition, (LMC), or regardless of feature size, (RFS). These concepts apply only to features of size such as holes, pins, shafts, slots, and tabs. MMC is the condition when the feature contains the most amount of material. For a hole, this means the smallest dimension allowed by the minimum tolerance. For a pin, this means the largest dimension allowed by the + tolerance.

⊕	$\varnothing$ 0.1 Ⓜ	A	B	C

An Ⓜ in a feature control frame as shown above (in the tolerance value compartment) means that if the feature is at MMC, this is the tolerance in its position. As the feature moves away from MMC (e.g. the hole gets larger), then there is an additional tolerance permitted on the position.

The Ⓜ modifier is most frequently used with holes that will be used with bolts to fasten parts together. Intuitively, a larger hole can be farther from its nominal location and the bolt will still go through. Of the three options, Ⓜ, Ⓛ, and Ⓢ, it is the easiest to satisfy.

If the Ⓜ symbol were replaced by Ⓢ in the feature control frame above, it would mean that the tolerance on the location of this hole is always .1mm, regardless of the size of the hole. Of the three options, this is the strictest.

If more than one datum reference is used, different modifying symbols may be placed after each datum reference symbol.

In the case where the important tolerance is not on the part itself, but on the projection to a mating part, the modifying symbol Ⓟ is used.

The software contains a syntax checker to check for the proper syntax of the GD&T symbol construction. This checks for proper use of datum references for geometric characteristic symbols that require them, and proper use of modifying symbols.

GD&T control frames will be placed on drawings that are made of the part. They will also remain with the part for other downstream uses, such as manufacturing and inspection.

Model Views

 Model views described on this page only available in I-DEAS 8 or above.

The 3D annotation described in the previous pages is 2D information living in a 3D world. Most of these annotation symbols only make sense when viewed from a good viewing angle. Also, when all of this annotation is placed on a part, the display can get very cluttered. The concept of model views on the 3D model in I-DEAS is a practical solution to these problems.

In a similar way that symbols are placed in the most appropriate view in traditional 2D drawings, model views can be used to store particular view directions and display annotation in the best views on a 3D part.

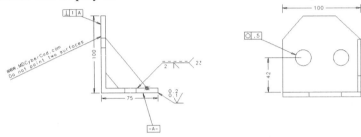

In the pictures above, the annotation has been organized into different model views. When the model view is selected, the view rotates as shown and only the specific annotation is shown. The tutorials to follow will show the specific instructions for creating and activating model views.

Although model views may be used as a 3D electronic alternative to 2D paper drawings, they can also be used to create 2D drawing views.

Part and Assembly Model Views

Model views can be created one at a time explicitly, or they can be created in sets, much like creating a standard four-view drawing. They can also be organized into folders of model views. You may use folders to organize views for different purposes, for example to create a folder for manufacturing inspection showing critical dimensions and annotation.

Model views can be created on both parts and assemblies. Assembly model views are much the same as part model views, with the addition of the assembly configuration. Each configuration stores its own model views.

 You can save steps by creating model views first, then creating annotation. By working in a model view, the plane of the annotation text is already defined.

2D Drafting

A typical process for creating 2D drawings is

1. get a 3D part,
2. create the 2D drawing and,
3. add annotation to it.

The order of these steps may be different, since annotation as described earlier in this chapter can be documented on the 3D part before creating the drawing. This order may depend on the practices of the company and the roles of the individuals involved.

The drawing can be stored electronically for retrieval, or plotted on paper.

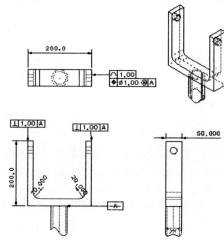

In I-DEAS 8, drawings are stored and managed in the model file, like parts and assemblies. This is different from earlier versions, where drawings were stored in external files. External files can still be used by exporting and importing drawings.

If you are using I-DEAS 8 or higher, skip the next page.

Drafting Setup Task *(Pre I-DEAS 8 Versions Only)*

 Note: This page applies only to versions of I-DEAS before I-DEAS 8. Skip this page if you are using the latest version, in which Drafting has been integrated as a task in the Design application.

Drafting Setup is a task in the Design application. First create parts and assemblies, then use Drafting Setup to create a drawing layout. The drawing may then be passed to the Drafting application for final detailing.

 The *Create_Layout* icon opens a form to enter the information to create a drawing of parts or assemblies. On this form, you will enter the drawing name, size, and select the part or assembly geometry to use. You can either select predefined layouts (such as a standard four view drawing), or create and place each view manually.

 After the drawing layout is defined, you can add new views and modify view scale and location. Icons are available to add more standard views of the geometry such as front, right, top, back, etc.

 Another stack of icons creates special views from existing views, such as section views, detail views, and auxiliary views.

 The second row of icons lets you move, resize, and change scale factors of views. Note that the *Move* command moves views, where a similar icon below moves dimensions and other annotations.

 You can add dimensions and notes to the drawing. You may also hide dimensions that are not needed on the drawing.

 The *Detailing* icon will bring the drawing into the Drafting application.

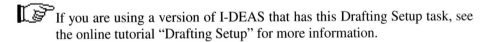 If you are using a version of I-DEAS that has this Drafting Setup task, see the online tutorial "Drafting Setup" for more information.

The information covered in the rest of this chapter describing the integrated Drafting task is similar to the Drafting Detailing application in your version except for the process of creating drawings from 3D parts and assemblies. See the online tutorials delivered with your version of the software for specific instructions.

Drafting Task

Drawings are created in the Drafting task of the Design application. Since this task only manipulates 2D information, there are a few differences in the Drafting interface from the user interface presented in earlier chapters.

You will first notice that the icon panel has a different appearance, with a "Command Option Area" in the center. This area will prompt you for actions and present menu choices and text entry fields.

Icons are contained in "stacks" as in other applications, but they can be used two ways. If you click the left mouse button once on an icon, the icon will change color, and you will be placed in that icon's mode for one command only. This mode of operation is called "spring loaded." If you click twice on an icon, it will turn gray, and you will be "locked" in that icon's mode of operation until you cancel out.

In the Drafting task, the left and middle mouse buttons work as previously defined to pick items and as a return key. The right mouse button is *not* the same as elsewhere in I-DEAS. The right mouse button in Drafting is used to cancel a command.

The Drafting task works with standard or custom drawing sizes.

Standard Drawing Sizes			
Size	Dimensions		
A-H	8.5	X	11 In
A V	11	X	8.5 In
B	11	X	17 In
C	17	X	22 In
D	22	X	34 In
E	34	X	44 In
A4-H	297	X	210 MM
A4-V	210	X	297 MM
A3	297	X	420 MM
A2	410	X	594 MM
A1	594	X	841 MM
A0	841	X	1189 MM

Creating a Drawing

 To start the drawing creation process, select the *Create Drawing* icon. In the Command Option Area, enter the drawing name, number, and select the bin of the model file where the drawing is to be stored. You must also select a drawing size using either metric or inch sizes

There are three ways to start the drawing-

> *Predefined View Layout,*
> *Create/Place Each View,* and
> *No Views.*

The first option allows you to pick from a set of predefined views, such as to create a standard four-view drawing. The second option lets you define and place views as you create the drawing. To create a drawing with no views is a way to create a drawing from scratch without using an associated 3D model.

To complete the first step of defining the drawing, click *Done* in the Command Option Area or click the middle mouse button.

If you are creating standard predefined views or creating and placing each view, you next select whether you want to use an associated 3D model in the views. The last part or assembly used will be the default selection, or you can select a different model by clicking on the folder next to the model name.

There are three ways to define what is meant by "front" for your views. The default is to click *Done*, and let the software choose the global XY view as the front. The second method is to use the 3D graphics viewer to pick a face and an "up" direction vector. If you are creating and placing each view, a third option is available to use model views.

When prompted, place the views by picking two corners and selecting *Done*.

The next step is to define the view scales. Using associated models, the scale can be inferred by the view size automatically.

If you are placing views manually, you will be prompted to place more views. By default, the option *Standard View Quick Mode* is on. This automatically infers which view to create by its placement relative to the front view. For example, if you place the next view above the front view, it is assumed to be a top view.

After selecting *Done*, the prompt area will ask you to select entities or a command, which indicates that you have finished the drawing creation process.

Drafting Views

A drawing contains views, each of which can contain unassociated 2D drafting entities or associated views of 3D models (parts or assemblies). By default, the first view is the main view which is as large as the drawing. It is recommended that you only draw the drawing border and title block in the main view. Drawing entities should be placed in views you create.

To work in a view, that view must first be made active either by using the *Workview* command, or by just clicking in it. The active view will have a dashed border. Most commands can only make changes to entities in this active work view.

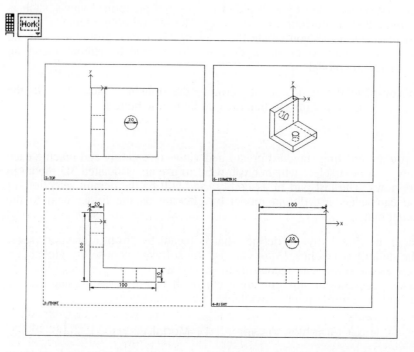

The hidden line display in each view displaying an associated model is processed using a coarse, medium, fine, or precise display option. By default, views are calculated using the coarse selection for speed of creating views. You can change the properties for any or all views using the *View Properties* icon and updating the drawing with the *Update* icon. Before exporting or plotting, you will normally want to use a precise display, especially if your drawing contains curved lines or arcs.

This area of the icon panel contains other icons for creating and defining views.

Display Commands

At the bottom of the icon panel are the commands for *Zoom All, Zoom Window,* and *Redisplay.* These commands allow you to zoom in on one view, part of a view, or on the entire drawing.

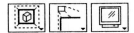

Dynamic Viewing function keys F1 and F2 can also be used to manipulate the display as in other tasks to pan and zoom. Because the Drafting task only works with 2D drawings, rotation controls are not used.

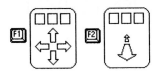

Creating 2D Geometry

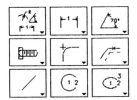

The icons just above the Command Option Area are the main entity creation commands. The icon in the lower left is the line creation icon. This stack also contains icons to create rectangles and polygons. The next two icons in the bottom row are used to create circles, arcs, and ellipses. The icons in the center row create symbols, fillets, chamfers, offset curves, and projections. Icons in the top row create dimensions.

An icon in the left stack of the bottom row is used to create variational geometry constraints on sections, similar to the variational geometry constraints in the Modeler task of the Design application, covered in Chapter 4.

Location Options

While you are creating lines or curves, a set of "location options" will be presented in this Command Option Area. There are options such as to turn the Dynamic Navigator on and off, to control reference axes, and to choose methods of explicitly locating points if you don't want to use the Dynamic Navigator. The icon on the left below controls the Dynamic Navigator. Users feel more comfortable with the concept of the Dynamic Navigator when they know they can turn it off if they find a case where it makes the wrong choice.

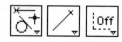

The second icon above offers a choice of how to manually specify digitized points. For example, the default method as indicated in the icon is to pick the end of a line. Other choices are available such as: to pick the center of the line, a percent along a line, a digitized location on a line, tangent to a curve, or at the intersections of lines.

Also with the location options are some status fields that let you numerically locate points rather than digitize them with the mouse. The fields X and Y will let you enter the absolute x and y locations of a point. Fields H and V let you enter the incremental x and y location of a point relative to the last point. To enter a point using polar coordinates, there are fields A for angle, and L for length.

The icon on the right, above, turns on *Auto Alignment*. With this option on, the cursor will "lock" on entities found in horizontal or vertical alignment with the cursor location. For example, when creating viewport borders, this feature makes it easy to create the borders exactly lining up with other viewports.

Undo/Redo Command

At the very bottom left of the icon panel are the commands *Undo* and *Redo*. If you make any mistake and want to back up, use the *Undo* command. If you undo too far, you can go back again with *Redo*.

Dynamic Navigator

The Dynamic Navigator works similarly as elsewhere in I-DEAS to give you graphic feedback, in advance, of what will be created and which entities will be selected. Different commands use the Dynamic Navigator function differently, based on the selections available for the specific operation.   When selecting geometry to delete or edit, the cursor will highlight the entities you are near before you click the mouse, to show you what will be selected.   When you are constructing geometry, the cursor will change shape to preview exactly what will happen when you click the mouse button.   For example, if you are near the end of a line, the cursor will show the symbol ×— to indicate the end of the line will be selected if you click the mouse at that location.   The table below summarizes the "preview cursors" used by the Dynamic Navigator to indicate what point will be selected.

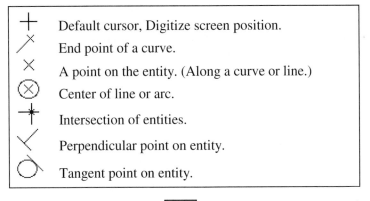

+	Default cursor, Digitize screen position.
✗	End point of a curve.
✕	A point on the entity. (Along a curve or line.)
⊗	Center of line or arc.
⊣	Intersection of entities.
⤬	Perpendicular point on entity.
⟳	Tangent point on entity.

Grid

An icon in the upper right corner of the icon panel gives you options pertaining to a grid that can be used to let you create points and line end points falling only on incremental grid locations.

Annotation Commands

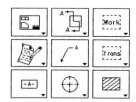

The top section of icons contains icons to create notes; feature control symbols and labels; circle centers, break marks, and cut planes. The icon in the lower right corner creates cross-hatching.

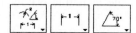

There are two ways to create dimensions. The icon on the left end of the row shown above creates linear, angular, and radial dimensions, using the Dynamic Navigator to choose which option you want. The shape of the preview cursor will change to indicate the type of dimension being created, as shown in the table below. If you have a lot of very small, closely spaced geometry, and want to manually create the different types of dimensions, then use other icons to explicitly create each type of dimension.

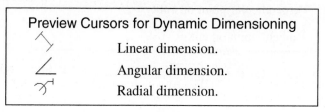

Editing Commands

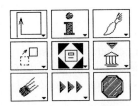

The icon at the upper left is the *Edit Entity* icon, to make changes to entities previously created. The next icon is for listing information about graphical entities. The icon in the upper right is to edit graphic attributes.

The left icon stack in the middle row contains icons for translating, copying; rotating; reflecting, and scaling.

An important editing command is the *Delete* command, on the lower left. You can either select the entities (using the select icon previously described) and then click the *Delete* icon, or you can first click the *Delete* icon, and then select the items to delete, followed by Return. The next two icons on the bottom row are for trimming and dividing lines and curves.

Selection Options

The Command Option Area in the center of the icon panel can also display a set of "selection options" if you are not using a command that requires the "location options" described earlier. These icons look like this:

The icon on the left is the *Select* icon. This will put you in a mode of selecting items one at a time. For example, you could individually select some lines and then press the *Delete* icon to delete the selected lines.

The icon in the center is the *Chain Select* icon. This mode will let you pick one line, and the program will automatically select all the connected lines.

The icon on the right is the *Window Select* icon. This allows you to select items using a rectangle, with options of selecting what is completely in or partly in the window.

Attributes

The *Graphic Attributes* icon shown above is found just below the Command Option Area. It is used to set default line color, weight, and style. Line and curve entities you create have attributes of a layer (1 to 256); a color; a curve font (solid, dashed, etc.); and a line weight (thin, medium, thick). It is usually easier to set these attributes before you create drawing entities than to edit them later.

Creating Assembly Drawings

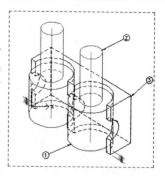

Assembly drawings are created by simply selecting an assembly instead of a part. One addition is that you must select an assembly configuration to display.

You may add a Bill of Materials to assembly drawings, with balloon callouts.

 To suppress instances in the assembly from the drawing, use *View Properties*, Hidden Line Processing Options (HLP), *Advanced Process, Suppress Instances, Instance List.* In versions before I-DEAS 8, suppress the instances in the assembly configuration.

Storing Drawings

When you created the drawing, you gave it a name in the model file bin. To save the drawing, save the contents of the model file with the *File, Save* command. You may also export the drawing in various formats such as IGES and DXF using the *File, Export* command. The drawing may also be checked into a library using the *Check In* icon.

 In versions before I-DEAS 8, drawings are written to the operating system as binary or ASCII files. An additional item for the drawing layout created by the Drafting Setup task is also stored in the model file.

Summary

This chapter has given an overview of drafting and annotation. The following tutorials and workshop will illustrate how to automatically create four-view drawings of parts made in previous workshops.

Where To Go For More Information

For more information on Drafting, see the manuals and online tutorials.

Help, Help Library
  Drafting User's Guide
  Drafting Programmer's Guide

☼ PARTING_SUGGESTIONS

Drafting Task (I-DEAS 8 or higher)

-Name your parts or assemblies before creating a drawing in Drafting.

-When you check a drawing into a library that is associated to a part, you will also check-in the part.

-Create views before creating geometry, rather than creating geometry in the main view. This will make it easier to modify this geometry later. Also, geometry in the main view will be displayed on every sheet of a multiple sheet drawing.

-Use a coarse Hidden Line Processing (HLP) display to create drawings. Convert to a precise display when you are ready to plot the drawing.

ϙ̓ PARTING_SUGGESTIONS

Drafting Setup Task (Pre I-DEAS 8 only)

-Remember to name your parts or assemblies before using Drafting Setup.

-When you check a part into a library that has an associated drawing setup, you will also have to check-in the drawing setup.

-If you work with very large drawings, there are some preferences that you should change.

Drafting Application (Pre I-DEAS 8 only)

-Drafting does not write directly to the model file. It uses its own drawing files.

-I-DEAS Data Management assigns an item name to track drawing files.

-Detailed drawings created from parts will have three items tracked by I-DEAS Data Management– the part, the drafting setup, and the detailed drawing file. The drafting setup and the detailed drawing file will be treated as one entity.

-After you create a detailed drawing from a drafting setup, when you get back into the Modeler task or the Drafting Setup task, check the drawing into a library.

-When working as a team, don't check out the part unless you need to change it. To modify annotation or add views, check out the drawing setup into your model file, which will automatically check out the drawing file as well. To actually change the part dimensions, you need to check out the part as well.

-After you exit Drafting, you should save the model file to permanently save the data management items that track the associativity between parts, drawing setup, and detailed drawings.

-Get in the habit of creating viewports before creating geometry, rather than creating geometry in the overall paper viewport. This will make it easier to modify this geometry later.

-There is an optional icon panel you can use that configures the icon panel to look more like the icons and the arrangement of icons used in the Modeler task. This panel is accessed using the pull-down menu Options, Modeler Layout.

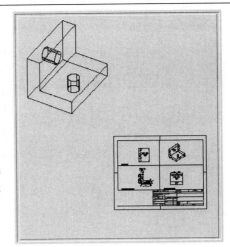

Tutorial: Drafting Setup

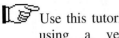Use this tutorial ONLY if you are using a version earlier than I-DEAS 8.

Before you start:

1. Skip this and the next tutorial if you are using I-DEAS 8 or higher.

2. You should already know how to create simple parts and manage them in a model file.

What you should learn:

1. How to create a Drafting Setup to create an associated 2D drawing of a 3D part.

2. How to bring this Drafting Setup into Detail Drafting to add annotation.

Tutorial Instructions.

Start I-DEAS with a new model file.

Follow the instructions in the tutorial below from the online Help Library.

Help, Help Library, Tutorials
Design - Part Modeling
1- Fundamental Skills
Drafting Setup

After completing this tutorial, continue with the tutorials on the next page.

 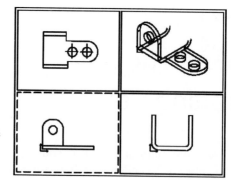

Tutorial: Drafting Detailing

 Use this tutorial ONLY if you are using a version prior to I-DEAS 8.

 Before you start:

1. Skip to the next tutorial if you are using I-DEAS 8 or higher.

 What you should learn:

1. How to create and manipulate drawing entities.

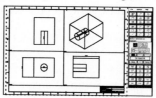

 Tutorial Instructions.

Whether you want to learn Drafting Detailing to add annotation to 2D drawings of parts or to create drawings from scratch, you will want to complete the series of tutorials from the online Help Library listed below.

Help, Help Library, Tutorials
  Drafting
    1- Fundamental Skills
        Introducing the Drafting Interface
        Creating Geometry: Lines Options
        Creating Geometry: Lines Rectangles
        Creating Geometry: Curves and Fillets
        Modifying Geometry
        Graphic Attributes
        Creating Dimensions
        Editing Dimensions
        Annotation
    2- Advanced Skills
        Views and Multisheets
        Projection
        Symbols and Crosshatching
        Plotting
        Translating Between Other Systems

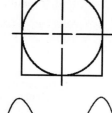

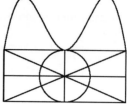

The remainder of the tutorials apply only to I-DEAS 8 or higher. Skip to the Workshop after completing the tutorials above.

 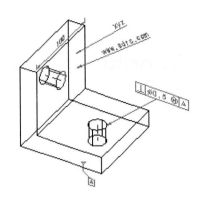

Tutorial: 3D Part Annotation

This tutorial demonstrates how to add annotation directly to 3D parts.

Before you start:

1. You must be using I-DEAS 8 or higher for this tutorial.

2. You should know how to create parts.

3. You should know how to manage parts in the bin.

What you should learn:

1. How to add notes to 3D parts.

2. How to add GD&T annotation to parts.

3. How to control the visibility of 3D annotation.

Tutorial Instructions.

Start I-DEAS with a new model file.

Follow the instructions in the tutorial below from the online Help Library.

> *Help, Help Library, Tutorials*
> > *Drafting*
> > > *3D Annotation*
> > > > *3D Part Annotation*

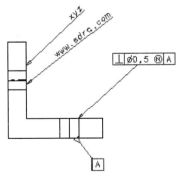

Save your model file for the next tutorials.

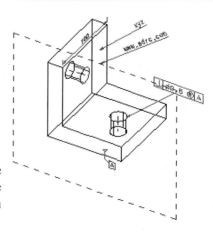

Tutorial: Using Part Model Views

This tutorial shows how to use part model views to organize and display part annotation on a 3D part.

Before you start:

1. You must be using I-DEAS 8 or higher for this tutorial.
2. You should know how to create annotation on a 3D part.

What you should learn:

1. How to create model views.
2. How to display annotation in particular model views.
3. How to activate model views.

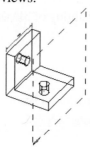

Tutorial Instructions.

Start I-DEAS with the model file from the previous tutorial, or create a simple part with some 3D annotation.

Follow the instructions in this tutorial from the online Help Library.

> Help, Help Library, Tutorials
> > Drafting
> > > 3D Annotation
> > > > Using Part Model Views

When you finish, save your model file for the tutorial *Creating Drawing Views from Model Views*.

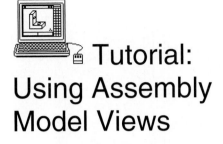

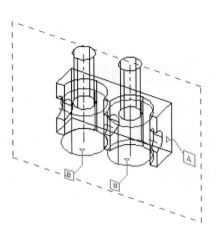

Tutorial: Using Assembly Model Views

This tutorial demonstrates how to use model views with an assembly.

Before you start:

1. You must be using I-DEAS 8 or higher for this tutorial.

2. You should know how to create parts and assemblies.

3. You should know how to use model views with parts.

What you should learn:

1. How to display part annotation on an assembly, and how to add annotation to assembly instances.

2. How to create and manipulate assembly model views.

3. How to organize assembly annotation in model views.

Tutorial Instructions.

Create the assembly in the tutorial below, or create a similar assembly on your own.

Design, Assemblies
    Fundamentals,
        Creating Assemblies

Follow the instructions in the tutorial below to add annotation and model views to this assembly.

Help, Help Library, Tutorials
    Drafting
        3D Annotation
            Using Assembly Model Views

Save this model file for the later tutorial *Creating Assembly Drawings*.

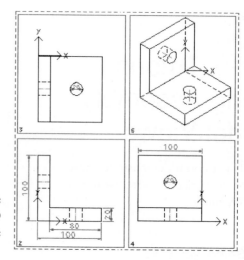

Tutorial: Creating Associative Drawings

This tutorial introduces the process of creating 2D drawings with views that are associative to 3D parts.

Before you start:

1. You must be using I-DEAS 8 or higher for this tutorial.

2. You should know how to create and manage parts.

What you should learn:

1. How to create a standard four-view drawing of a part.

2. How to save the drawing.

3. How to modify the part and update the drawing.

Tutorial Instructions.

Either open a new model file to create the part as instructed in the setup steps of the tutorial, or open a file containing a part of your own.

Follow the instructions in the tutorial below to create the simple part shown, and then to create a standard four-view drawing of it.

Help, Help Library, Tutorials
  Drafting
    Associative Drawings
      Creating Associative Drawings

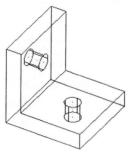

When finished, save your model file containing this drawing. You may want to use it in one of the tutorials to follow.

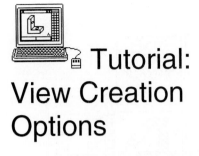 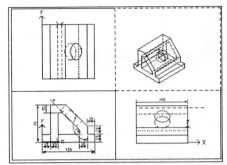

Tutorial: View Creation Options

This tutorial shows other ways to create associative views, and some important view options such as the hidden line processing resolution.

Before you start:

1. You must be using I-DEAS 8 or higher for this tutorial.
2. You should know how to create and manage parts.
3. You should know how to create a standard associative drawing.

What you should learn:

1. How to create drawing views using
   Create/Place Each View,
   3D Model Window, and
   Quick Mode.
2. How to modify view borders and view display processing options.
3. How to update a drawing, and recognize out-of-date views.

Tutorial Instructions.

Follow the instructions in the tutorials below, starting with a new model file.

> *Help, Help Library, Tutorials*
> > *Drafting*
> > > *Associative Drawings*
> > > > *View Creation Options*

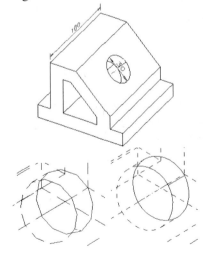

You will create the part on the right and the drawing shown at the top of the page.

The effect of changing from the fast hidden line processing to a precise display is shown when you zoom in on the hole in the isometric view.

Save your model file for the next tutorial.

Tutorial:
Adding
Section, Detail, &
Auxiliary Views

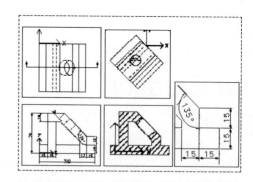

This tutorial shows the types of views that are derived from geometry in a parent view.

Before you start:

1. You must be using I-DEAS 8 or higher for this tutorial.
2. You should know how to create associative drawings.
3. You should know how to change view options.

What you should learn:

1. How to create section, auxiliary, and detail views.

Tutorial Instructions.

Start I-DEAS with the model file containing the part from the previous tutorial.

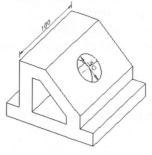

Follow the instructions in the tutorials below.

Help, Help Library, Tutorials
Drafting
Associative Drawings
Adding Section, Detail and Auxiliary Views

You will create a new drawing and add section, auxiliary, and detail views.

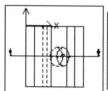

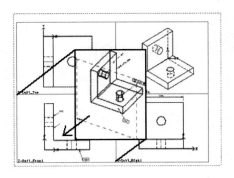

Tutorial: Creating Drawing Views from Model Views

This tutorial shows how to create drawing views directly from the model views stored with a part.

Before you start:

1. You must be using I-DEAS 8 or higher for this tutorial.
2. You should know how to create part model views containing 3D part annotation.

What you should learn:

1. How to create drawing views from part model views.
2. How to modify the scale, origin, and border of views.

Tutorial Instructions.

Start I-DEAS with the model file saved from the earlier tutorial, *Using Part Model Views.* This file contains a part with model views.

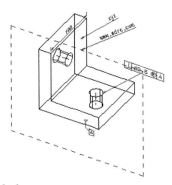

Follow the instructions in the tutorial below.

Help, Help Library, Tutorials
Drafting
Associative Drawings
Creating Drawing Views from Model Views

In this tutorial, you will create views by selecting and placing the predefined model views on the drawing.

 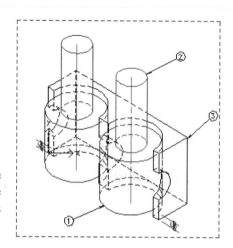

Tutorial: Creating Assembly Drawings

Assembly drawings are created the same as part drawings, with the additional concepts of configurations and adding a Bill of Materials.

Before you start:

1. You must be using I-DEAS 8or higher for this tutorial.
2. You should know how to create associative drawings of parts.
3. You should know how to create assemblies and use configurations.

What you should learn:

1. How to create an assembly drawing.
2. How to add a Bill of Materials to the drawing.

Tutorial Instructions.

Start with the model file saved in the earlier tutorial *Using Assembly Model Views*, or recreate the assembly in the *Creating Assemblies* tutorial.

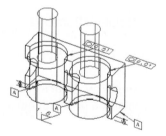

> Help, Help Library, Tutorials
> Design, Assemblies
> Fundamentals
> Creating Assemblies

Annotation from the model views tutorial is not required. This annotation will be hidden from the drawing in the tutorial to follow.

Follow the instructions in the tutorial below.

> Drafting
> Associative Drawings
> Creating Assembly Drawings

In this tutorial, you will create a drawing of the assembly in the design configuration and add a Bill of Materials. When you are finished, you may delete this model file.

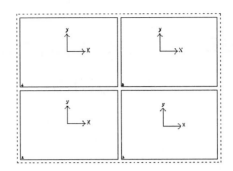

Tutorial: Creating Unassociative 2D Drawings

This series of tutorials covers all of the basic techniques for creating unassociative geometry. These techniques can be used either to create complete drawings from scratch, or to add geometry and annotation to associative drawings.

Before you start:

1. You must be using I-DEAS 8 or higher for this tutorial.
2. You should know how to start and exit I-DEAS.
3. You should know the basics of the user interface.

What you should learn:

1. How to create empty drawing views.
2. How to create 2D drawing geometry and annotation.
3. How to modify drawing geometry and views.

Tutorial Instructions.

Start I-DEAS with a new model file.

Follow the instructions in the tutorials listed below.

> *Help, Help Library, Tutorials,*
> *Drafting*
> *Unassociative Drawings*
> *Creating Unassociative 2D Drawings*
> *Creating Geometry: Lines & Options*
> *Creating Geometry: Lines & Rectangles*
> *Creating Geometry: Curves and Fillets*
> *Modifying Geometry*
> *Graphic Attributes*
> *Creating Dimensions*
> *Editing Dimensions*
> *Annotation*
> *Using Multisheets*
> *Projection*
> *Symbols and Crosshatching*
> *Translating Between Other Systems*

Workshop 7A:
Manufacturing Tolerance

The pump we have been designing in earlier chapters does not require particularly tight tolerances. It does, however, have some fabrication and installation tolerance considerations. The Pivot Support must be installed parallel to the well casing so that the Stroke Slide slides smoothly on the Guide Pipe. Also, the two bearing holes in the Pivot Block must be parallel or one of the Offset Links will carry all of the load, causing faster wear of the bearings. The Pivot Support width at the top should match the Pivot Block. If it is too tight, the pivot won't move. If it is too loose, the Pivot Block could slide back and forth, causing the Stroke slide to be pulled up at an angle, which again will cause faster wear.

This workshop will demonstrate the analysis of some of these tolerances, using the *Tolerance Analysis* icons.

Since tolerance analysis was not covered in any of the previous tutorials, this will be new material, not a review like many other workshops.

Before you start:

   1. For this workshop you will need the part "Pivot Support" created in Workshop 2A at the end of Chapter 2.

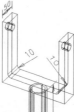

   2. You should know how to create and manage parts.
   3. You should know how to use constraints with wireframe geometry.

After you're done, you should be able to:

   1. Add tolerance values to dimensions.
   2. Analyze tolerance variation using a tolerance model.

Workshop Instructions:

Start the Design application, the Modeler task with a model file containing the parts from Workshop 2A. You may either use the same model file, or use a new file and import the parts from an external file or library. If you saved the program file from that workshop, you may run that program file to recreate the parts.

 If you have not completed Workshop 2A, go back and complete the five pages where these parts are created.

If you are using a version of I-DEAS with libraries, *reference* the Pivot Support part out of the Pump Parts library. *(Reference the parts because the modifications made in this workshop will not be checked back into the library.)* Convert the references into copies without notification of changes. *(This will allow you to make modifications, but will break all association to the original library parts.)*

Get the Pivot Support part out on the workbench.

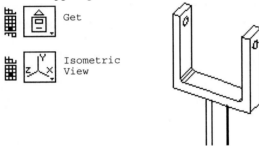

Get

Isometric
View

One type of tolerance is to add an upper and a lower bound tolerance on any dimension. Click twice on any edge of the Pivot Support to pre-select the <u>part</u>. The part should then be highlighted with white corner boxes. Use the *Modify* command to show the dimensions of the pivot support.

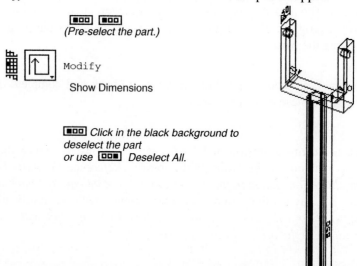

[■□□] [■□□]
(Pre-select the part.)

Modify

Show Dimensions

[■□□] *Click in the black background to deselect the part or use* [□□■] *Deselect All.*

Modify the 850 mm vertical extrude dimension to add a tolerance value. Specify that this length can have a tolerance of +/- 10mm. If the text size of dimensions is too large or small, use the *Appearance* icon and resize the text with the *Autoscale* button.

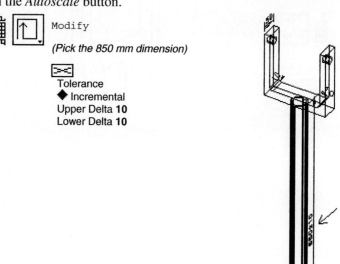

Modify

(Pick the 850 mm dimension)

[>─<]
Tolerance
◆ Incremental
Upper Delta **10**
Lower Delta **10**

Zoom in on the upper end of the Pivot Support. (Your part may have different dimensions than shown.)

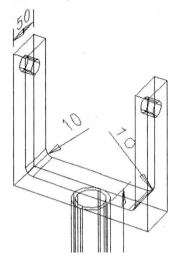

Extract the wireframe section from the U-shaped extrude <u>feature</u> of this part. (Don't extract the wireframe boundary of a face of the part, or it will not contain the dimensions and constraints used to create the feature.)

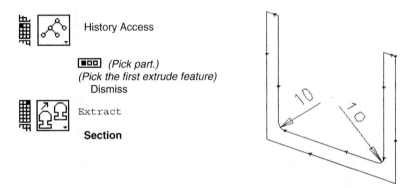

History Access

         (Pick part.)
     (Pick the first extrude feature)
         Dismiss

`Extract`

Section

Put the Pivot Support part away, leaving just the wireframe section shown above on the workbench.

 `Put Away`

Delete all of the dimensions except the two radius dimensions shown above.

 `Delete`

Ground the point in the lower left corner.

Constrain & Dimension

Anchor

Add vertical and horizontal point-to-point dimensions as shown, picking the corner at each end of the dimension. When you create dimensions for a tolerance model, the order of picking each end of the dimension makes a subtle difference. You should pick the end on the fixed (the datum side) first, and then the movable end. If you do it the other way around, the program will give you a non-fatal warning when you analyze the tolerance model.

Dimension

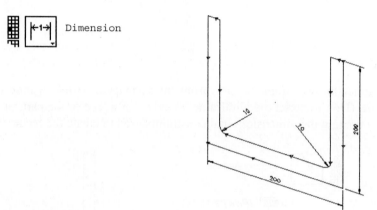

Add point-to-point dimensions across the sides at the top. Modify the values of the dimensions and add tolerances as shown.

Dimension

Modify

Tolerance
◆ Incremental
Upper Delta **1**
Lower Delta **1**

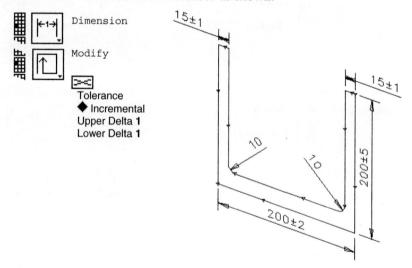

Add a <u>reference</u> dimension across the top of the bracket. A redundant dimension is automatically created as a reference dimension.

`Dimension`

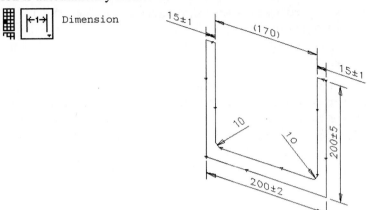

Add a tolerance to this reference dimension. Note that the tolerance value on a reference dimension is fundamentally different than a tolerance on a driving dimension. A tolerance on a driving dimension states how accurately this dimension can be manufactured. The tolerance on a reference dimension is a goal that we would like to achieve, based on the contribution of all the other driving dimensions. When we analyze the tolerance model, this tolerance will be the target, to see how many parts will statistically fall within the desired goal.

`Modify`

Tolerance
◆ Incremental
Upper Delta **3**
Lower Delta **3**

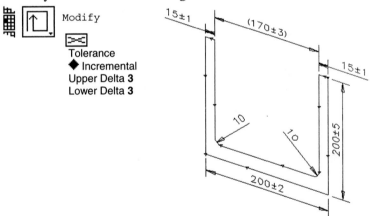

Pick this reference dimension as the driven dimension to be analyzed.

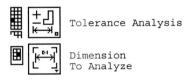

`Tolerance Analysis`

`Dimension`
`To Analyze`

Use the *Verify* command to verify that the tolerance model is ready to be analyzed. (It isn't.) What is wrong? The warnings say that the model is not fully constrained. Use the *Show Free* command to find out what DOFs are still unconstrained. (You may ignore the warnings suggesting that you could have used line to line instead of line to point dimensions.) Fully constrained lines are shown in blue.

Verify

(Pick any edge of the section.)

Constrain & Dimension

Show Free

(Pick any edge of the section.)

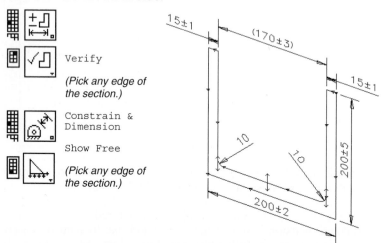

Add one more vertical dimension on the thickness of the bottom plate.

Dimension

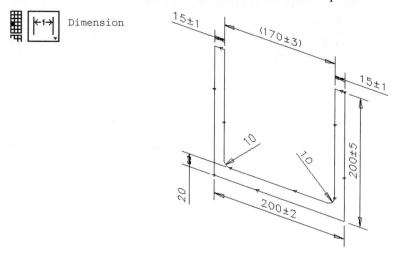

Check the free DOFs again. Depending on how you sketched the original section, you may also need to add other constraints, such as parallel, perpendicular, or collinear constraints.

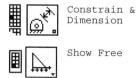

Constrain & Dimension

Show Free

Analyze the tolerance model for the reference dimension.

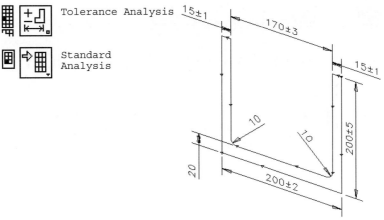

```
Tolerance Analysis Results:
---------------------------
Name:D17   (Your dimension label may be different.)
Nominal: 170.0
Stat. Upper Tolerance: +2.44949
Stat. Lower Tolerance:-2.44949
Upper Tolerance: +3.0
Lower Tolerance:-3.0
Worst Up_Stk:  +4.0
Worst Low_Stk: -4.0
Acceptance Rate: 99.975960
```

Which tolerance values contribute the most to the variation in the reference dimension? With the given tolerances on the driving dimensions, how many parts would fall within the requested tolerance on the reference dimension? Notice that the statistical acceptance rate is almost 100%, even though the worst case stack-up (4.0) would indicate a tolerance much higher than the goal of (3.0).

When you are finished, put away the tolerance model, naming it "Tolerance Model."

Save your model file for the next workshop.

Workshop 7B: Product Documentation

This workshop will give you a chance to review the topics presented in the preceding tutorials. You will add annotation and create a drawing of the Pivot Block to document the product design.

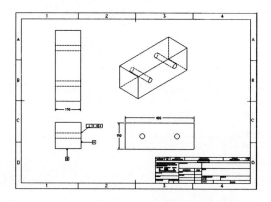

Before you start:

1. For this workshop you will need the part "Pivot Block" created in Workshop 2A at the end of Chapter 2.

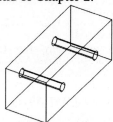

2. You should know how to create and manage parts.
3. You should know how to create drawings.

After you're done, you should be able to:

1. Add GD&T annotation to parts.
2. Create a drawing of a part.

Workshop Instructions:

Start the Design application, the Modeler task with the model file from the previous workshop containing the parts from Workshop 2A.

 If you have not completed Workshop 2A, go back and complete the two pages where this part is created.

Get the Pivot Block part out on the workbench.

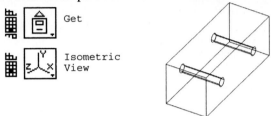

Get

Isometric View

Add a datum reference symbol to the side of the block, naming this datum reference surface "A". Depending on your version, the prompt sequence may be slightly different below.

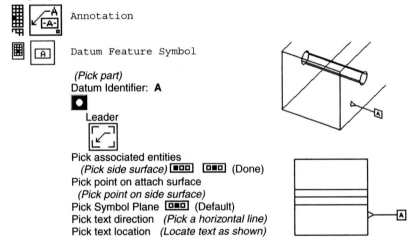

Annotation

Datum Feature Symbol

(Pick part)
Datum Identifier: **A**

Leader

Pick associated entities
  (Pick side surface) (Done)
Pick point on attach surface
  (Pick point on side surface)
Pick Symbol Plane (Default)
Pick text direction *(Pick a horizontal line)*
Pick text location *(Locate text as shown)*

If you prefer to use the ANSI standards, modify this GD&T Datum symbol to conform to the ANSI format. You may also set this as the default for future annotation symbols created.

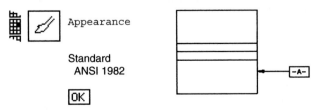

Appearance

Standard
  ANSI 1982

OK

Add a second datum reference symbol to the bottom of the block, naming this datum reference surface "B".

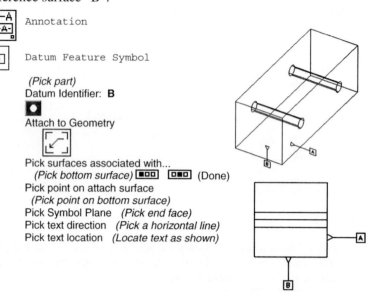

Annotation

Datum Feature Symbol

(Pick part)
Datum Identifier: **B**

Attach to Geometry

Pick surfaces associated with...
  (Pick bottom surface) ■□□ □■□ *(Done)*
Pick point on attach surface
  (Pick point on bottom surface)
Pick Symbol Plane   *(Pick end face)*
Pick text direction   *(Pick a horizontal line)*
Pick text location   *(Locate text as shown)*

Add a GD&T Feature Control Frame to specify that the hole is perpendicular to within a tolerance of one degree to datum A.

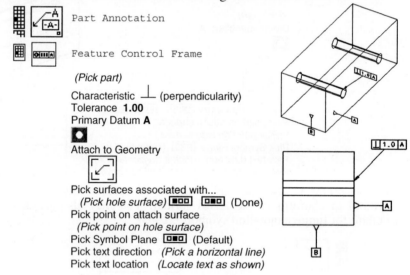

Part Annotation

Feature Control Frame

(Pick part)

Characteristic ⊥ (perpendicularity)
Tolerance **1.00**
Primary Datum **A**

Attach to Geometry

Pick surfaces associated with...
  (Pick hole surface) ■□□ □■□ *(Done)*
Pick point on attach surface
  (Pick point on hole surface)
Pick Symbol Plane □■□ *(Default)*
Pick text direction   *(Pick a horizontal line)*
Pick text location   *(Locate text as shown)*

Display the dimensions of the part.

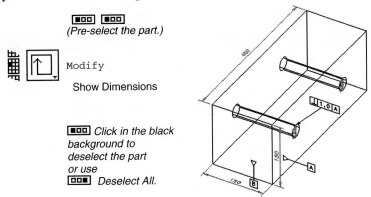

(Pre-select the part.)

Modify

Show Dimensions

Click in the black background to deselect the part or use Deselect All.

Add tolerance information to the block dimensions. Allow the length dimension to vary +/-5, and the other two dimensions to vary by +/- 2mm.

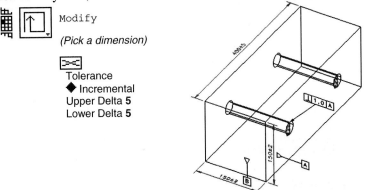

Modify

(Pick a dimension)

Tolerance
◆ Incremental
Upper Delta **5**
Lower Delta **5**

If you are using I-DEAS 8 or higher, convert the three major block dimensions to document dimensions. They will remain displayed after the update. They will also be identified as document dimensions when you create a drawing of the part. (Skip this step if you are using the Student Edition 1.0 or a version before I-DEAS 8.)

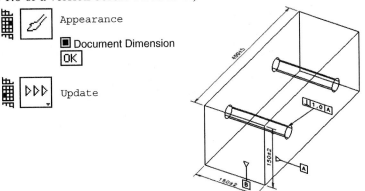

Appearance

■ Document Dimension
OK

Update

Save your model file.

Practice creating a drawing of the Pivot Block part, using what you learned in the tutorials.

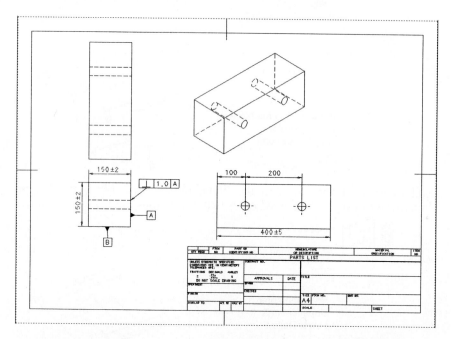

If you need additional help on your specific version, refer to the online tutorials described earlier.

You may delete your model file when you are finished. This file is not used in any other tutorials or workshops. Keep the pump parts in case you want to use them to practice making drawings, or for use in later chapters describing other applications in I-DEAS.

Chapter 8
Manufacturing

NC (Numerical Control) Machining 236
Axes of Motion 237
NC Machine codes...248
Manufacturing Application 239
Part Geometry...239
Generative Machining Terminology.......................240
Machining Sequences 241
Setup..241
Post Processing...241
Summary 242
Where To Go For More Information 242
Parting Suggestions 242
Tutorial: Introduction to Generative Machining 243
Tutorial: Building a Setup Assembly 244
Tutorial: Milling Projects 245
Tutorial: Turning Projects 246
Workshop 8: NC Machining..247

NC (Numerical Control) Machining

Possibly the most important "downstream" use of the solid model geometry is to manufacture the part. The whole reason for the design and analysis described so far in this Student Guide is that we eventually want to make physical parts. The manufacture of most parts will either use some form of machining process to make the part from a larger piece of material, or will use a machining process to make molds to form the part using some other manufacturing process such as plastic injection molding. This chapter will discuss the steps required to take a part model, as shown in earlier chapters, and use the Manufacturing application to define the tool operations to machine the part. While there are important advantages to designing parts using solid modeling, and to manufacturing them using NC (Numerical Control) machining, the advantages that result when using the combination of the two are even greater.

An integrated use of solid modeling and NC machining results in more accurate parts produced in less time. Since parts can be made directly from the edges and surfaces taken from the solid model, this geometry does not have to be manually interpreted by a Machinist or re-entered by an NC Part Programmer. Since computers are used for both the design and the machining, the information can be transferred electronically, without the need for paper drawings or even an intermediate media such as magnetic tape or punched paper tape. This eliminates duplication of work and reduces the chances for error, resulting in a synergistic combination of solid modeling and NC machining.

The use of the combination of solid modeling and NC machining also gives more creative freedom in the design. If a designer knows ahead of time that NC machining will be used to make the part, the design will not have to be constrained to simple machine operations that can be done by hand. Once the metal stock is mounted in a fixture on the machine, it doesn't cost any more for the machine to cut esthetic curves rather than straight lines, or to make extra cutouts in the center to save weight.

Axes of Motion

Several different types of machining operations can be performed by NC machines, such as milling, lathe turning, flame water-jet or laser cutting, nibbling, etc. In each of these machines, the motion of the cutter relative to the part is described using a "right hand rule" rectangular cartesian coordinate system as has been used throughout this Guide. In some types of machines the cutter will move, and in others, the part will move and the cutter will remain stationary. This makes no difference when creating the NC program, since the tool motions will be described relative to the part in either case.

These motions are designated by convention as X, Y, and Z. The general convention for machining is that the Z axis is parallel to the axis of spindle rotation, which is often vertical for a milling machine or horizontal for a lathe. The X axis is usually the longest axis perpendicular to this Z axis. Some machines will have more than these three axes, such as rotating the spindle or the work piece in one or two directions. Five axis machines or more are common.

One of the first considerations in planning the machining process is deciding how many axes of machining are required, and what machine should be used. The number of axes of the machining operation is the number of independent motions that are required. For example, for flame cutting parts out of sheet metal, only two axes, X and Y, are required. Milling machines often have three axes of motion, an X and Y motion of the table, and a Z motion of the rotating spindle. For many parts however, the type of machining that is required is called two and a half axis machining. This refers to the fact that only two axes are ever changed simultaneously. The third axis may be used to drill holes or to set the depth of the spindle for pocketing, profiling, or plunge cutting, but not at the same time as X and Y motion. Although the result is a three dimensional part, the machining operation used is called two and a half axis instead of three axis. Three axis machining is where all three axes are controlled at the same time, such as to machine a sculpted surface.

It is important to understand the difference between two and a half axis (2.5D) or three axis, (or any multi-axis machining), since the geometry preparation will be different. In the first case, we will only need the projected 2D profile of the part. For multi-axis machining, we will need to be concerned with the surfaces of the part as well as the wireframe edges.

NC Machine codes

The actual commands for a specific machine are usually made up from a sequence of letters and numbers. These commands range from simple auxiliary functions such as selecting tools or waiting for a manual tool change, to basic move commands to move the spindle to cut material, to more complex "canned cycles" such as drilling a sequence of holes to the same depth at several locations. The first character in a command is usually the letter "N" followed by a line number, or sequence number, that allows the program to be listed and read by humans. The next letter on many machines is a G, followed by a number that tells the computer what type of operation to perform, such as to turn the tool spindle on and off, to set machining feed rates, or to move to a given X, Y, and Z location. These instructions are not universal between machines, although the formats look similar. The following is a typical section of an NC program.

```
N0005   G00 X-0.25 Y1.5 Z3.0
N0010   Z0.1
N0015   G01 Z-.375 F400
N0020   X-0.1562 Y1.4459
N0025   G03 X-0.125 Y1.4375 I0.0312 J0.0541
N0030   G01 Y2.625
N0035   G02 X0.375 Y3.0625 I0.4375 J0.0
N0040   G01 X2.625
```

Because these specific machine instructions are different for each machine, most NC programs initially write a file called a "cutter location" file, which is a list of the motions of the center of the cutter in a standard format. This cutter location file is converted to specific machine instructions like those above using a "post processor" written for the specific machine.

NC machines today are controlled by a computer, so these machines are commonly called CNC, for Computer Numerically Controlled machining. (In this chapter, when we use the term NC, this can be assumed to mean the same thing as CNC.) The use of the computer control has added flexibility to the NC program, such as not having to worry about a rigid format for the instructions. Computer control gives the operator the ability to make local changes to the program, such as to edit some of the lines in the program, or to make small offsets to the tool paths to adjust for changes in tool length or diameter due to tool wear. It is important to note that the instructions in the NC program are for the motions of the center of the cutter. What would happen if the tool wore down a little bit due to wear and re-sharpening? If the change in radius of the cutter was not taken into account, the program would machine a part slightly larger than it should be.

Manufacturing Application

In the Manufacturing application, the following tasks are available:

Manufacturing Application Tasks

Modeler

Generative Machining

Assemble Setup

The Modeler task is the same as described in the Design section of this Guide. The Assemble Setup task is basically the same as the Assembly task, with some specific additions to understand specific manufacturing terminology such as fixtures and clamps.

Part Geometry

The first step is to create either a solid model of the part in the Modeler task in any one of the I-DEAS applications. Remember that the part must be named before you can use it in the Generative Machining task.

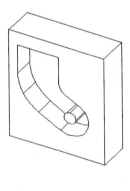
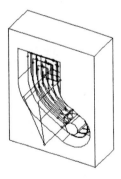

Generative Machining Terminology

A manufacturing "job" is associated with the part to be machined, just like drawings and finite element models are associated to the part. Multiple jobs can be created from the same part, just like multiple finite element models can be created for a part to study different effects.

A job can contain multiple "setups." A setup is identical to an assembly, as discussed in Chapter 6. The only difference is that specific terminology is used in the setup assembly to refer to some of the assembly instances as fixtures, clamps, and the machine. A job will contain multiple setups if you need to perform another operation where the part, clamps, and fixtures need to be rearranged.

Each setup can contain multiple OpGroups, which are groups of machining operations. It is somewhat arbitrary how you organize operations into OpGroups, but typically one OpGroup would be used for each feature, such as a pocket. All of the operations to machine that pocket, such as a rough cut then a profile operation, would be grouped in one OpGroup.

The relation of the part to the machining operations is shown in outline form below.

Part	The part to be machined must be created and named in the Modeler task.
Job	A job represents your entire job plan. It is the collection of the setups and operations required to machine a part or assembly of parts.
Setup	A setup is an assembly of the part and optionally: stock material, fixtures and clamps. Setup1 is automatically created. When you select the command Add Setup, a second assembly is created. Use the Assemble Setup task to modify the orientation of the Part and clamps.
OpGroup	A group of machining operations. Generally, use a new OpGroup for each feature of the Part.
Operation	Operations are steps like profiling, volume clear, and copy-mill (surfacing).

 The shown outline above is versions before I-DEAS 8. In I-DEAS 8 or higher, a job is not stored as a "child" of a part. This allows a job to be associated with more than one part.

Machining Sequences

The actual sequence of machining operations can be re-ordered by the program according to rules defining efficient machining, such as to group like operations and tools together, and to perform operations with larger diameter tools first.

Setup

A setup is an assembly containing the part to be machined plus other parts representing things like the machine, fixtures, and clamps. A setup is automatically created for the part you are machining. If you want to change anything about this setup, you may use the Assemble Setup task. This task is very similar to the Assembly task in the Design application (Chapter 6). The main difference is that this task has a stack of icons to directly build machining Setup assemblies, such as to add clamps, fixtures, the machine, and stock to the Setup.

Including stock is optional in the setup assembly. With the toggle "Calculate In-Process Stock" turned on, the program will perform the cuts on this stock part to monitor the actual machining results. This in-process stock passes to the next Setup in the same Job, to use to position fixtures and clamps. The part is used to calculate toolpaths, but material is removed from the stock. The stock is also used to determine tool travel speed, based on whether the tool is cutting material from the stock or not.

  In versions prior to I-DEAS 8, the stock and part are grouped together as a subsystem called a "Workpiece."

In I-DEAS 8 or higher, both part and stock instances are placed in the top level of the setup assembly.

Other instances in the Setup assembly such as the machine or fixtures can also be single instances or subsystems. The program will automatically avoid clamps. The Add Stock to Setup command is used to add stock to the machining assembly Setup.

 Add Stock
to Setup

Post Processing

The Post Processor takes the Cutter Location File from the machining operations and writes a Tape File. The Post Processor will need a machine file as input to describe how to convert cutter motions into specific machine commands for the machine you will use.

Summary

This chapter has outlined some of the main concepts for the Manufacturing application. This chapter has mainly covered milling. For more detailed information, especially on topics such as turning and using machining coordinate systems, read the online manuals in the Help Library.

Where To Go For More Information

For the complete description on using the Manufacturing application, see the online Help Library manuals.

> *Help, Help Library*
> > *Manufacturing User's Guide*
> > > *Generative Machining*
> > > *Assemble Setup*
> > > *C-Post*
> > > *G-Post*

☼ PARTING_SUGGESTIONS

-When you get parts to machine out of a library, you should Reference them instead of Check Out, so that others can access them at the same time. If you need to change the part to suppress features or to add geometry, you can change your library status to Copy.

-If you need to simply rotate a part, instead of converting the library status from a reference to a copy, you should rotate the instance in the setup assembly. In versions prior to I-DEAS 8, rotate the workpiece assembly to rotate both the part and the stock.

-If you have trouble picking the entities you want on the display, set the *Filter* to define the types of entities you want to pick.

-Before you leave the Manufacturing application, Put Away the Setup in the Assemble Setup task.

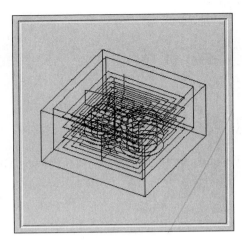

Tutorial: Introduction to Generative Machining

This tutorial demonstrates the basics of creating an NC job using Generative Machining.

Before you start:

1. You should know how to create and manage parts.
2. If you need more modeling practice, see the tutorials in the list:

    Help, Help Library, Tutorials
       Manufacturing
          Modeling Fundamentals

What you should learn:

1. How to create an NC job.
2. How to add operations.
3. How to animate the toolpath.

Tutorial Instructions.

Start I-DEAS with a new model file and follow the instructions in the tutorial below, found in the online Help Library.

Help, Help Library, Tutorials
   Manufacturing
      Milling Projects,
         Introduction to Generative Machining

In this tutorial, you will create a part and then create the toolpaths to machine it.

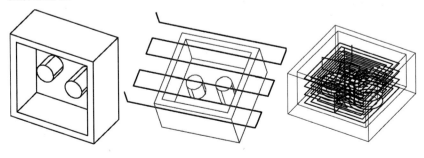

 Tutorial:
Building a
Setup Assembly

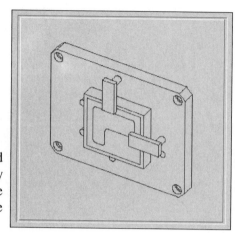

Adding fixtures, stock, and clamps to the setup assembly models the whole machine setup and generates more accurate toolpaths.

Before you start:

1. You should know the basics of creating an NC job with Generative Machining.

What you should learn:

1. How to add stock, fixtures, and clamps to the setup assembly.

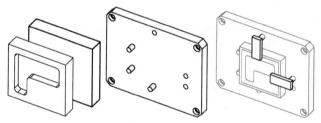

Tutorial Instructions.

Start I-DEAS with a new model file and follow the instructions in the tutorial below. If you are using the Student Edition 1.0, be sure to read the special instructions, which is first in the list of tutorials.

Help, Help Library, Tutorials
Manufacturing
Milling Projects,
Building a Setup Assembly

Continue with the next tutorial to see the in-process stock and the effect of the interfering clamp on the toolpath.

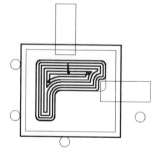

Generating In-process
Stock and Checking Validity

Tutorial: Milling Projects

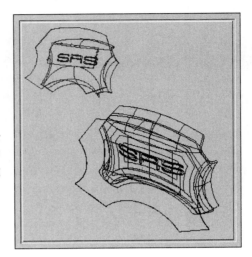

There are about a dozen other tutorials available that show different aspects of milling operations.

Before you start:

1. You should know the basics of creating an NC job and a setup assembly with Generative Machining.

What you should learn:

1. How to use tools and tool catalogs.
2. How to machine holes.
3. How to use face mill and volume clear operations.
4. How to use profile, copy mill, and flowline operations.
5. How to use manual milling operations.
6. How to use multiple setups and indexing.
7. How to create CL files and post process the output.

Tutorial Instructions.

Follow the list of tutorials in the Milling Projects list. If you are using the Student Edition 1.0, be sure to read the special instructions, which is the first item in the tutorial list.

Help, Help Library, Tutorials
Manufacturing
Milling Projects

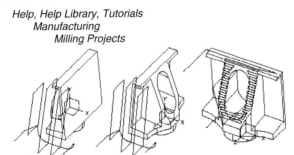

When you finish the milling tutorials, there are more tutorials in the online Help Library on turning operations.

Tutorial: Turning Projects

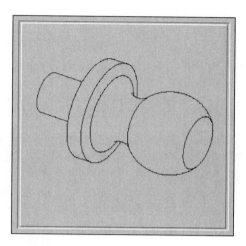

Turning operations share many of the same concepts as milling operations. Machine and stock instances are required in the setup assembly. In addition, the tools and types of operations are different than milling.

Before you start:

1. You should know the basics of creating an NC job and a setup assembly with Generative Machining.

2. It will help to first understand the basics of milling operations.

What you should learn:

1. How to create a turning setup.

2. How to use turning operations including rough turn, cutoff, thread, and groove operations.

3. How to create turning tools.

Tutorial Instructions.

Follow the list of tutorials in the Turning Projects list. Be sure to read the special instructions if you are using the Student Edition 1.0.

Help, Help Library, Tutorials or *Help, Tutorials,*
Manufacturing
Turning Projects

Workshop 8:
NC Machining

In this workshop, we will create the NC tool paths to machine a mold half for the "Tee" part as modeled in Workshop 2B.

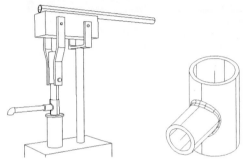

We will make a mold for the outer shape of one side of the Tee, so we won't need to model the complete part; we will only model the outer surface, as shown below.

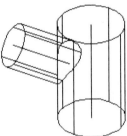

Before you start:

1. You should know how to create and manage parts.

2. You should complete at least the first Manufacturing tutorial first.

After you're done, you should be able to:

1. Create NC tool paths for 3D milling operations.

 Some of the steps in the workshop to follow will deviate slightly from the printed instructions, depending on your version of the software. Be sure to read the prompts carefully.

Workshop Instructions:

Enter the Manufacturing application, the Modeler task with a new model file.

    $ `ideas` <RETURN>

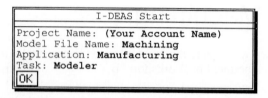

Get a primitive cylinder from the Parts Catalog with a Radius of 20 mm and a Height of 80 mm. Preselect the part by clicking twice. (White boxes should box in the cylinder.) Translate this cylinder 40 mm in the Y direction, and then rotate 90 deg about the Z axis (through a point at the origin).

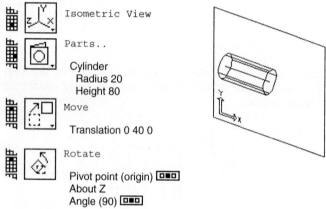

    Isometric View

    Parts..

      Cylinder
        Radius 20
        Height 80

    Move

      Translation 0 40 0

    Rotate

      Pivot point (origin) ▢■▢
      About Z
      Angle (90) ▢■▢

Get another primitive cylinder with a radius of 30 mm and a Height of 100 mm. Join the two together with relations turned off.

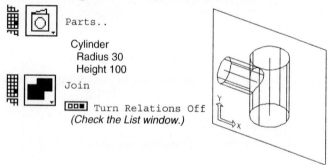

    Parts..

      Cylinder
        Radius 30
        Height 100

    Join

      ▢▢■ Turn Relations Off
      (Check the List window.)

Give this part the name "T Surface".

    Name Parts

      T Surface

Sketch a rectangle on the workplane larger than the part.

 Rectangle by 2 Corners

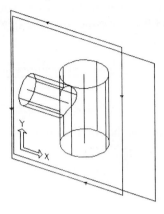

Extrude this rectangle 50 mm into a new part to become the Mold Half. Name this part "Mold Half."

 Extrude

 Flip Direction

 Name Parts

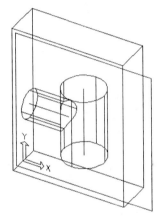

Cut the Tee from the Mold Half. (Make the cut with Relations off.)

 Cut

Pick T Surface
Pick Mold Half

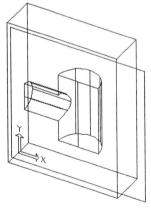

Switch to the Generative Machining task. In the Student Edition 1.0 and versions before I-DEAS 8, you will be prompted for the part to machine and the job name. In I-DEAS 8 or later, use the icons *Create Job* and *Add Part*.

>   Pick part to machine- **Mold Half**
>   Job Name: **JOB1**

Bring up the Job Planning form. Modify the OpGroup to add an operation.

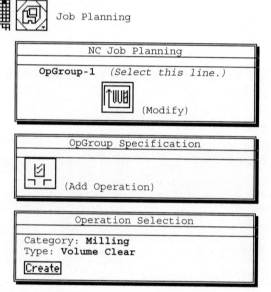

```
             NC Job Planning

OpGroup-1   (Select this line.)

                [icon]
                        (Modify)
```

```
           OpGroup Specification

     [icon]  (Add Operation)
```

```
          Operation Selection

Category: Milling
Type: Volume Clear
[Create]
```

"Walk" through the icons to define the Operation. First pick the geometry to machine.

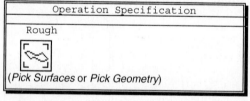

```
          Operation Specification

Rough

  [icon]

(Pick Surfaces or Pick Geometry)
```

(Use Shift-Click to pick the two cylindrical surfaces.)

[OK]

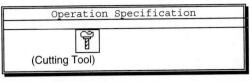 Define the tool geometry.

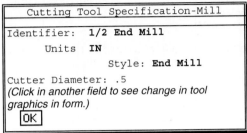

Examine some of the machining parameters. You don't need to change anything.

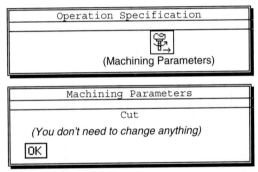

Process the tool path for this operation.

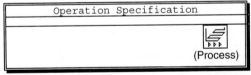

Notice: Since you did not add a stock part or define the limits of the stock, your toolpath may be different for the top level of cut. For more information, search for the word "stock" in the Manufacturing: Generative Machining User's Guide in the online Help Library.

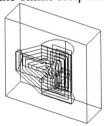

Display the Mold Half shaded, and animate the tool motion.

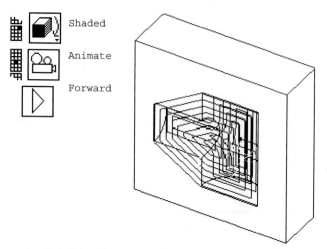

Shaded

Animate

Forward

Bring up the Job Planning Form. Modify the OpGroup to add a second operation. This time, we will do a "copy mill" operation to follow the cylindrical surface.

Job Planning

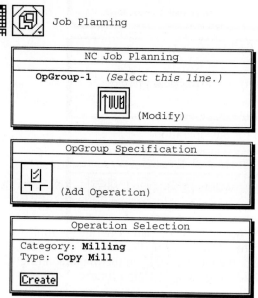

"Walk" through the icons to define the Operation. First, pick the geometry to machine.

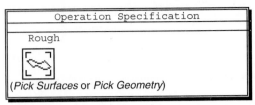

If your version of the software displays a Surface Selection form, press the icon Pick Surfaces.

Otherwise, just use Shift-Click to pick the two cylindrical surfaces.

Define the tool geometry.

Examine some of the machining parameters. You don't need to change anything.

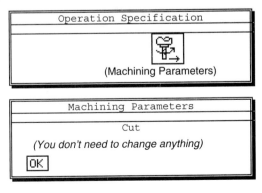

Process the tool path for this operation.

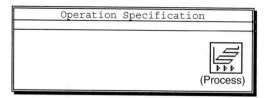

Animate the tool motion.

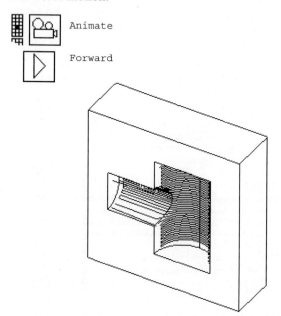

Animate

Forward

When you are finished, put away the Setup Assembly.

 Assemble Setup

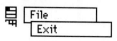 Put Away

(You may turn off the toolpath display using *Display Filters*.)

Exit when you are finished.

File
Exit

Chapter 9 Simulation

Finite Element Analysis

Finite Element Analysis (FEA) is a process which can predict deflection and stress on a structure. (It can also be used to calculate heat transfer, fluid flow, or electrical fields, etc. but the introductory discussion here will be confined to structural analysis.) Finite Element Modeling (FEM) divides the structure into a grid of "elements" which form a model of the real structure. Each of the elements is a simple shape (such as a square or a triangle) for which the finite element program has information to write the governing equations in the form of a stiffness matrix. The unknowns for each element are the displacements at the "node" points, which are the points at which the elements are connected. The finite element program will assemble the stiffness matrices for these simple elements together to form the global stiffness matrix for the entire model. This stiffness matrix is solved for the unknown displacements, given the known forces and boundary conditions. From the displacements at the nodes, the stresses in each element can then be calculated.

A finite element is derived by assuming a form of the equation for the internal strains. Some elements are defined to assume that the strain is a constant throughout the element, while others use higher-order functions. Using these equations and the actual geometry of a given element, the equilibrium equations between the external forces and the nodal displacements can be written. There will be one equation for each degree of freedom for each node of the element. These equations are most conveniently written in matrix form for use in a computer algorithm. The matrix of the coefficients becomes a "stiffness matrix" that relates forces to displacements.

$$\{F\} = [K] * \{d\}$$

Even though the unknowns are at discrete degrees of freedom, the internal equations were written for strain functions that represent a continuum. This means that even though the finite element model has a discrete number of equations, if the right elements are chosen, it is possible to converge on the correct answer with a less-than-infinite number of nodes and elements.

A finite element model is the complete idealization of the entire structural problem, including the node locations, the elements, physical and material properties, loads and boundary conditions. The model will be defined differently for different types of analyses: static structural loads, dynamics, or thermal analysis.

A finite element model is often made of more than one element type. The finite element model is made to mathematically model the deflection of the structure, not to look like it. Parts of a structure might be best modeled with beam elements, and other parts with thin shell elements.

The accuracy of the resulting solution will depend on how well the structure was modeled, the assumptions made for loads and boundary conditions, and the accuracy of the elements used for the given problem. In general, the solution will be more accurate as the model is subdivided into smaller elements. The only sure way to know if you have sufficiently converged on the final solution is to make more models with finer grids of elements and check the convergence of the solution.

New finite element users often make the mistake of thinking the goal of making a finite element model is to make a model that looks like the structure. The purpose of finite element modeling is to make a model that behaves mathematically like the structure you are modeling, not necessarily one that looks like the real structure. An experienced user learns how to choose the right type of element and how fine to make the grid in different areas of the model.

An often-overlooked source of errors in a model is wrong assumptions made in loading and boundary conditions. It is too easy to trust the answers from a finite element model to the last decimal place, and to forget the crude assumptions made for the loading and boundary conditions. If there is a question about how to model boundary conditions, use your model to test the sensitivity to the different methods rather than just following one path and getting one answer. It has been said: "The purpose of analysis is insight, not numbers."

Steps in Finite Element Analysis

Finite element modeling consists of three steps. These are:

PRE-PROCESSING

SOLUTION

POST-PROCESSING

Pre-processing includes the entire process of developing the geometry of a finite element model, entering physical and material properties, describing the boundary conditions and loads, and checking the model.

The solution phase can be performed in the Model Solution task of the Simulation application, or in an external finite element analysis program. Model Solution can solve linear and non-linear statics, dynamics, buckling, conduction heat transfer, and potential flow analysis. For other types of analyses, the finite element model information can be written in the format required for an external finite element solver such as NASTRAN, ANSYS, or ABAQUS.

Post-processing involves plotting deflections and stresses, and comparing these results with failure criteria imposed on the design such as maximum deflection allowed, the material static and fatigue strengths, etc. If we only wanted to know if the part would survive the load, all we would need to see would be a yes or no answer. This is usually not the case. We would like to be able to see the results in different display formats, which will give us insight into why the part will fail and how to improve the design. There are two questions that must be answered in post-processing: "Is the model accurate?" and "Is the structure satisfactory?"

There are many possible sources of error in your model, such as the coarseness of the finite element grid, the type of elements used, or incorrect material properties. This is why post-processing should include checking for errors that might not have been detected while building the model. One basic check you should always perform is some kind of hand calculation to make sure you have not made any gross errors such as a slip of a decimal point in entering material properties. It is also recommended to plot displacements before looking at stresses, because displacements usually make more intuitive sense than stress. Make sure that the shape of the deflection looks right before going on. Often errors made in boundary conditions can be detected by a close look at the deflection shape, such as points that should move but don't, or the wrong slope at the restrained points. Make sure your model is free from errors before you make judgments on the structure you are modeling.

Simulation Application

There are several tasks in the Simulation application covering the three steps of pre-processing the model, solution, and post-processing. The basic tasks used for the three steps of finite element modeling include:

Simulation Tasks

Pre-Processing
    Modeler
    Meshing
    Boundary Conditions

Solution
    Model Solution

Post Processing
    Post Processing

Although nodes and elements can be created manually in the Meshing task, finite element models are more easily built by automatically meshing parts created in the Modeler task, either in the Simulation application or the Design application. An advantage of using I-DEAS as an integrated MCAE package is that the solid model geometry can easily be shared between applications.

The Meshing task is used to create nodes and elements, check the model, and enter material and physical properties. The Boundary Conditions task is used to apply boundary conditions to the part geometry or the nodes and elements. Either meshing or boundary conditions may be done first, depending on the type of boundary conditions. When the model is complete, it is solved in the Model Solution task and the results displayed in the Post-Processing task.

Other tasks are available for laminate analysis, beam section modeling, adaptive meshing, optimization, and dynamic analysis.

I-DEAS Test, to be covered in Chapter 10, also shares much in common with the Simulation application. The same finite element nodes and elements are used to display dynamic mode shapes measured by the process of modal testing on a physical structure. Modal testing is a very effective way to check the accuracy of a finite element model, especially if that model will be used for dynamic analysis.

Many of the commands and terms used in Simulation are the same as in other parts of I-DEAS. It is assumed that you already understand the concepts presented in earlier chapters. Additional terminology will be added to these basics in the next few chapters.

"Nodes" are coordinate points in 3D space where "elements" will be connected, loads will be applied, boundary restraints imposed, and displacement information will be computed. In other finite element programs, nodes are sometimes referred to as "grid points." Each node has up to six degrees of freedom, depending on the element types connected to it. Nodes are a geometric entity in the database that can be displayed or not; controlled by the Display Filter switches. A label displaying the node number can also be turned on and off.

An analysis will be performed using a "solution set" which contains a "boundary condition" set of "restraints" and "loads."

To help simplify the creation of a model and aid in displaying the results, named "groups" of nodes and elements can be formed. Groups can be used to display or process a subset of the model.

Analysis results are stored in the model file as "analysis datasets." Each dataset contains one set of displacements or stresses that can be displayed. Stresses are stored in a dataset as the entire stress tensor, so any stress characteristic can be computed and displayed within the Post-Processing task.

Element Types

One of the most important choices in making a finite element model is which element type to use. Most finite element programs include a "library" of element types to choose from. Elements are categorized by family, order, and topology.

The element family refers to the characteristics of geometry and displacement that the element models. In general, you should use the simplest element type that will model your problem. It is also important to consider what type of output you need.

The most common families used for typical structural models are: Beam, Plane Stress, Axisymmetric Solid, Thin Shell, and Solid. Beam elements can be used to make a very efficient finite element model to predict overall deflection and bending moments but will not be able to predict the local stress concentrations at the point of application of a load or at joints. Thin Shell elements can be effectively used for structures with relatively thin walls such as molded plastic or sheet metal parts where bending and in-plane forces are important. However, this type of element will not be able to predict stresses that vary through the thickness of the stress due to local loading effects. The most general element is the Solid family, but this type of model will need more elements to get the same accuracy for bending applications where a thin shell could have been used. A finite element model using solid elements may look the most realistic, but simpler elements may facilitate more efficient calculations when used appropriately. For example, the Offset Link shown below from Workshop 2 could be modeled using solid elements, thin-shell elements (two ways), or beam elements, depending on the desired result.

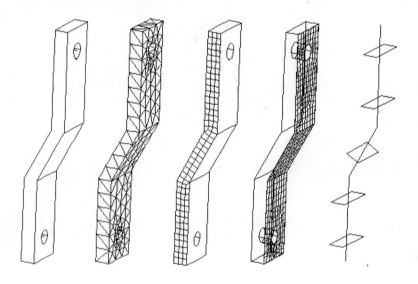

Elements are also categorized by order. The order of an element refers to the order of the equations used to interpolate the strain between nodes, such as linear, parabolic, or cubic. Linear elements have two nodes along each edge, parabolic have three, and cubic have four.

Element topology refers to the general shape of the element, such as triangular or quadrilateral. The topologies depend on the family of the element. In general, you should choose quadrilateral over triangular elements for structural models, since the quadrilateral element, having more degrees of freedom, can more accurately match the true displacement function.

Elements also contain different numbers of degrees of freedom at each node. For example, some elements are intended to model only two-dimensional problems, and may only contain two degrees of freedom per node. In general, the maximum number of degrees of freedom at each node is six: three translations and three rotations, although not all elements will use all six.

I-DEAS Element Types

Family	Order	Topology
Lumped Mass Spring Damper Gap Constraint Rigid Rigid Surface		
Rod Beam Pipe Plastic	Linear Parabolic	
Axisymmetric Shell Membrane Plane Stress Plane Strain Thin Shell Thick Shell Axisymmetric Solid	Linear Parabolic Cubic	Quad Triangle
Solid	Linear Parabolic Cubic	Wedge Brick Tetrahedron

Managing Finite Element Models

Finite element models are stored in the same bins in the model file as parts. A finite element model is associated with a part, just like drawings. For example, after you have created a drawing and two different finite element models of a part, the *Manage Bins* form will look something like this:

```
Main Bin
    Part1
        Drawing 1
        FE Model 1
        FE Model 2
```

If the part is modified, the finite element models can be updated to reflect the changes. If the part is deleted, all of the associated drawings and finite element models are also deleted.

Part Geometry For Finite Element Modeling

It would be rather tedious if all models had to be built using only the manual methods of creating nodes and elements discussed so far in this chapter. Finite element models can be automatically created on parts from the Modeler task. This chapter builds on the concepts from Chapters 1-5. It is assumed that you are familiar with this material. Although these tasks are the same in the Design and Simulation applications, this chapter will present additional commands and techniques that are used specifically for building models for finite element meshing.

Before discussing the ways geometry can be modeled, it will help to get an overview of the process of generating a finite element mesh on part geometry.

Nodes and elements can be automatically created on the edges, surfaces, or volumes of parts in the Meshing task. Beam elements can be generated on edges, thin-shelled elements on surfaces, or solid elements on volumes. Parts must be named before they can be used for meshing.

The following part was meshed with a thin-shell mesh on one surface to analyze the stress at the radius. Even though the part model was a 3D solid model, only a 2D finite element model was needed.

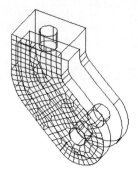

The meshing operation is a two-step process. First the parameters defining the size and type of elements and other attributes are defined on the particular edges, surfaces, or volumes of the part. The second step is to generate the mesh on this geometry.

The following icons are used to define the mesh parameters on part geometry. The first three icons are used to define meshing attributes for the geometry to be meshed. The most important meshing parameter is the element size. A global element size can be set on each edge, surface, or volume to be meshed. The last icon below is used to define local element sizes at particular points. Other parameters will be discussed in the next chapter, which focuses on the Meshing task.

	Define Shell Mesh	Set Thin-Shell meshing parameters on surfaces.
	Define Solid Mesh	Set Solid Element meshing parameters on part volumes.
	Define Beam Mesh	Set Beam Element meshing parameters on part edges.
	Define Free Local	Define local element lengths to override the global element length setting for free meshing

To preview the result of the mesh settings graphically before actually performing the meshing, these commands are used:

	Modify Mesh Preview	Preview element size before meshing.

The icon commands below are used to generate the mesh of nodes and elements with the current mesh settings that have been defined.

	Shell Mesh	Generate Thin-Shell meshes on part surfaces.
	Solid Mesh	Generate Solid Element meshes on part volumes.
	Beam Mesh	Generate Beam Element meshes on part edges.
	Mesh on Part	Generate elements on a part with multiple edges/surfaces/volumes defined.

Part Model Abstraction

Often, the model geometry we want for finite element modeling may be slightly different than the original 3D part model. As shown previously, an abstraction was made to create a mesh on only a 2D surface of a 3D model. This can often be done by just meshing one exterior surface. Sometimes you may need to generate a new cross section in order to simplify a 3D model to 2D. As shown below, if we want to perform a rough check on a 2D model to show the stress concentration around the radius in this "Tee" part from Workshop 2B, we could cut the part with a plane to generate a 2D surface through the cross section.

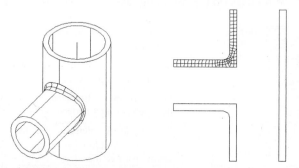

Another common extraction of solid models is to suppress or delete design details that we are not concerned with in the finite element model. For example, if we want to do a buckling analysis on the Offset Link part from Workshop 2B, we might not be interested in the holes. We could either delete them, or suppress them. To suppress a feature of a part keeps that step in the history of the part, but "turns it off" from the model we will use for finite element modeling. This option allows the Analyst to directly use a model created by the Designer. *Suppress* is an option under the *Modify* command when a feature is selected.

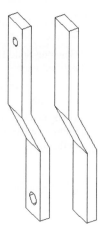

Non-Manifold Part Geometry

Sometimes finite element models will combine element types. For example, part of the finite element model may be modeled with solid elements, but thin-shelled elements may be used to model a thin rib. The data structure of part models allows geometry to be modeled this way.

A part model that has less than two or more than two surfaces meeting at any shared edge is called a "non-manifold" part. This can't occur in physical reality, but it is allowed in I-DEAS to create geometry for special cases such as a finite element with combined element types. Parts created in the Modeler task can contain solid volumes, open surfaces, and wireframe geometry. This allows you to create any special geometry for finite element meshing.

Another type of non-manifold part is a part that has been partitioned into multiple volumes.

A concept required for non-manifold geometry is that of the "material side." All of the surfaces of a manifold part have material only on the inner side. If you add a flange protruding from this solid using an open surface, it will have material on neither side. If you partition the part with a plane passing through the solid to cut it into two volumes, the partition plane has material on both sides.

A common finite element analysis situation is a model with more than one material. When you mesh a part volume, all the elements within that volume will be created with the material properties defined for that volume. In order to automatically generate a model with different material properties in different regions, the regions must be partitioned into different volumes. The command to do this is *Partition*, found in the Modeler task. This command will partition a part, using another part as a cutting tool.

 Partition

Surface Interpolation

In a thin-shelled finite element model, ideally the elements should lie on the mid-plane between the physical part surfaces being represented. Usually, it is a minor deviation to just model the elements on one of the surfaces, since the part must be thin by definition to allow the assumption of modeling it with a thin-shelled element. In some cases however, you may want to model the elements where they really belong, especially if the part is relatively thick, or if the two surfaces are not parallel. The command *Interpolate Surfaces* in the Modeler task will create a new surface interpolated between two others. The *Midsurface* command will create midsurfaces for a whole part.

Midsurface

Interpolate Surfaces

Surfacing Tools

Other surfacing tools are available, such as to cut a part with a plane cut to produce a cross section to create a 2D mesh. Some of the same surface creation tools discussed in Chapter 5 such as lofting, sweeping, and fitting surfaces to data points or boundaries may also be used to create or modify surfaces for meshing.

If you are starting from 2D or 3D wireframe geometry, the first step is to create surfaces for meshing. Depending on the type of surface, a number of commands can be used to create surfaces, such as *Mesh of Curves* or *Surface by Boundary*. Another convenient command to create trimmed surfaces inside a boundary of curves is the *Auto Trim* command, which defines boundary loops and generates surfaces inside the boundary.

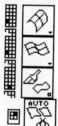

Mesh of Curves

Surface by Boundary

Surface Cleanup

Automatic Trim

Next, after creating surfaces, they need to be stitched together for meshing so that they share common edges. A good check to see where surfaces may be adjacent but are not stitched together is the command *Show Free Edges*.

Show Free Edges

Where there are free edges between surfaces, they need to be stitched together using the *Stitch* command.

Stitch Surface

Meshing Task

This section covers manual and automatic meshing to create nodes and elements using the Meshing task. Before we discuss automatic meshing methods, it will be useful to discuss manual creation options, to get a better foundation for automatic meshing.

Manual Node and Element Creation

Nodes can be created manually by keying in their coordinates or generated by copying, reflecting, or generating nodes between two other sets of nodes. Nodes are created by specifying the coordinates in any existing coordinate system. This is called the "definition" coordinate system for the node. Nodes also have another coordinate system called the "displacement" coordinate system. Unless you change it, the displacement coordinate system will be the global cartesian system. A displacement coordinate system other than the global cartesian system can be used by modifying the nodes after you have created them. This can be used to supply a boundary condition in a cylindrical coordinate system, for example.

Three kinds of coordinate systems are available in I-DEAS Simulation: cartesian, cylindrical, or spherical. These three coordinate systems are pre-defined; you can also define others. The global cylindrical coordinate system defines point locations with coordinates (R, Theta, Z). The global spherical coordinate system uses (R, Theta, Phi).

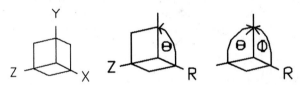

You can also create other local coordinate systems by translating or rotating one of these types. Coordinate systems can be created and then translated and rotated, or created by defining them with respect to an existing coordinate system.

When you create a mesh on a part, the coordinate system of the part is used. This will usually be aligned with the global coordinate system, but not always. For more information on part coordinate systems, search in the Finite Element Modeling User's Guide for the word "coordinate."

Elements can be manually created by picking nodes or by generating elements from existing elements. The normal manual method is to use the cursor to pick the nodes defining the element (four for a quadrilateral, etc.). In the "cursor menu" that appears when you are picking nodes, one of the options is to automatically pick the closest nodes, so you can just pick a screen location in the center of four nodes to create a quadrilateral element.

Elements can be created by copying or reflecting, like nodes. These operations will create nodes and elements at the same time. These copy and reflect operations create elements of the same type as the parent elements set. Other types of element generation such as extrude or revolve create elements of a different type. For example, you can extrude quadrilateral elements into 8-noded solid elements. This is a useful method to generate 8-noded solid brick elements without using mapped mesh volumes, which will be discussed below. A similar element generation method can create thin shells on the surface of solid elements by using the surface coating creation command.

Elements contain an element label, an element type, a list of nodes that make up the element, a color, a material property table ID, and a physical table ID. Any of these element attributes can be modified after the element is created, but the element type can't be modified to a type that uses a different number of nodes.

Meshing on Part Geometry

Automatic meshing uses part edges, surfaces and volumes. Parts do not need to be closed solid parts to use them for meshing, as shown in the last workshop. Solid elements can be generated on part volumes, thin-shell elements on part surfaces, or beam elements on part edges.

All three families of elements require that a material property be defined. It is often easier to create the material table before meshing, rather than modify the default property tables that the program creates for the elements if one does not exist. Solid and thin-shell elements may also require physical property tables, which also may be created before meshing. Beam elements also require one more property; a cross section definition. This topic will be discussed later in the chapter.

Mapped Mesh and Free Mesh

Nodes and elements are generated by one of two methods, mapped or free mesh. Each surface is defined as one or the other when defining mesh parameters.

Mapped meshing requires the same number of elements on opposite sides of the mesh area, and requires that mesh areas be bounded by three or four "edges." If you define a mapped mesh area with more than four curves, you must define which vertices are the topological corners of the mesh. The illustration below shows two different ways the corners could be defined for the same part geometry.

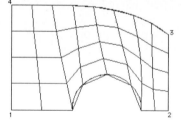 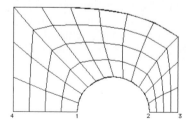

Mapped mesh boundaries with three corners will generate triangular elements in one "degenerate" corner.

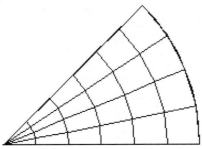

The mesh density is controlled by the number of elements per edge, and biasing of element size toward the end or the center of edges.

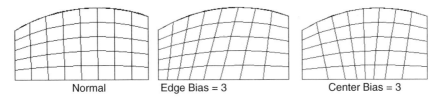

Normal   Edge Bias = 3   Center Bias = 3

Free meshing allows more flexibility in defining mesh areas. Free mesh boundaries can be much more complicated than mapped mesh areas without subdividing into multiple regions. The mesh will automatically be created by an algorithm which tries to minimize element distortion (deviation from a perfect square). Free mesh surfaces can easily have internal holes, where mapped mesh surfaces can't. The type of geometry shown below would have to be further subdivided to use mapped meshing.

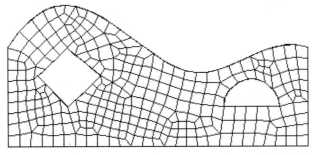

Free meshing is controlled by two parameters assigned to each mesh surface or volume that affect the size of the elements generated. The first is the element length, which is the nominal size of elements the program will attempt to generate. The second parameter controls mesh refinement at curves in the model by controlling how much deviation is allowed between straight element sides and curved boundaries. This parameter is expressed either as a percent deviation or an absolute number.

Mesh density can also be controlled by "Local Element Sizes" set on different vertex points on the boundary curves using the icon command *Define Free Local*. By varying these size settings, you can have substantial control over the mesh density in different areas. There may be times when you want more control over the mesh generation than the first two parameters provide. For example, you might want a finer mesh at one hole in the part, but not at all the holes. The parameter controlling curvature-based refinement would control all curved surfaces, but by assigning local densities, you can create smaller elements just where you want them.

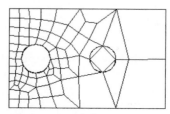

The meshes generated can be mapped onto the surface by one of two methods, found under Free Options. The default method is to map the mesh using "parametric space meshing." The equations for the surface are stored as functions of X, Y, and Z for the parameters s and t. Parametric space meshing develops the mesh in s and t coordinates and then maps it into three dimensional space. Parametric space meshing can't be used for surfaces where the parameters s and t converge at "poles" (cones and spheres). In these cases, switch to "maximum area plane." This method develops the mesh on a projected "shadow" of the surface on a flat plane. This method is more limiting in the maximum angle of curvature from one side of a mesh to the other. A practical limit may be 90 degrees, where with parametric space meshing, you can mesh a surface with a curvature of more than 180 degrees.

Curve and point geometry may have to be manipulated depending on the type of mesh you want (mapped or free), the location of applied loads, or other modeling considerations. Curves may be divided to add additional points. In the meshing process, nodes will be placed at the end of every curve. This may be used to further control mesh generation, if you want to specify node locations for loads or boundary conditions. Curves can be created on surfaces if areas need to be subdivided.

Avoid curves that are small relative to the element size. This can lead to errors generated by the program when trying to generate the nodes and elements. Also avoid curves and surfaces with tight curvatures compared with element sizes. If you really need to model these small details, you will need smaller elements. If small radii and features are not required in the finite element model, suppress them.

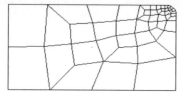

Mapped mesh on a volume is more restrictive than free mesh on a volume, since the mapped volume must be bounded by 5 or 6 surfaces. In the case of 6 surfaces, the interior is topologically a "box" and solid brick elements can be generated. If 5 surfaces enclose the volume, the volume is wedge shaped. Elements generated will be solid brick elements except at the last edge, where wedge elements will be used.

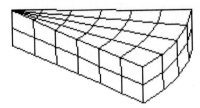 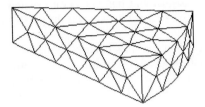

Some analysis programs may require either tetrahedral or brick elements. If this is the case, you may have no choice of which type of mesh volume to use, since free mesh volumes give tetrahedral elements and mapped mesh volumes produce brick or wedge elements.

Another method for creating solid brick elements is to generate a mesh or thin-shell elements on one surface, and then extrude these elements into solid brick elements.

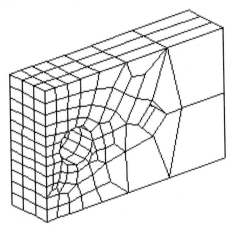

The software also has concept of "section meshing." This is a way of combining surfaces for meshing to create elements that are larger than the individual surfaces defining a part. This allows meshes to span across a section defined by multiple surfaces or regions. For more information on this topic, see the online tutorial "Section Meshing."

Material Properties and Physical Properties

Each element contains a material property ID which refers to a table of material properties. Every element must reference one material table. One table may be referenced by many elements. Material properties can be isotropic, orthotropic, or anisotropic. Material properties can be defined before creating elements or during the element creation process. If no material table has yet been created, the software will force you to create one. The default material properties represent steel.

If a material has been assigned to a part in the Modeler task, that material will be used by default for the finite element model. It makes sense then to assign a material right at the start on the part model, so that information is available to all of the "downstream" applications such as finite element modeling.

Physical properties are also referenced by elements. These represent factors like element thickness and beam cross-section properties. The default physical properties are usually meaningless and should not be used. Some elements do not need any additional physical properties, but the element must still reference a "dummy" physical table.

Groups

A group is a user-defined subset of the model. Groups may be used to simplify selection of elements or for display flexibility. Different areas or different materials in your model can be placed in different groups. Groups can be stored and retrieved. Groups can help others understand your model later if you form self-explanatory groups of different parts of your model. It is recommended that you create meaningful groups and store them while you are creating your model. This will also be very useful later when you want to plot results. It will be easiest to plot a stress contour on one part of your model by selecting one of these previously defined groups.

Model Checking

The Meshing task also contains several checks to help you identify modeling errors in your finite element model. Typical problems that can be checked are duplicate nodes, duplicate or missing elements, and highly distorted or warped elements.

One of these checks is an element free edge check. This check will plot the free edges of elements not connected to another element. This can be a very useful check in finding element connectivity problems. Normally, this will plot the outer boundary of the model, which is where the elements are not connected to others. If elements adjoin each other edge to edge but reference duplicate coincident nodes rather than share the same nodes, an extra line will show up in the free edge plot. This represents a "crack" in the model. Duplicate elements defined by the same nodes will cause neither element to be plotted in this check, and a missing line may show up in the plot.

Element distortion is another popular check. Values are reported by the distortion check from -1.0 to 1.0. A value of 1.0 represents a perfect square (a circle fits inside). Values less than 0.0 are horrible. A typical rule of thumb is that values should be between .4 and 1.0, but there is no exact cut-off for what is not acceptable. It depends on the type of analysis to be performed and where the badly distorted elements are located in the model. Avoid highly-distorted elements in important areas such as high stress locations. Sometimes, due to the geometry you are modeling, distorted elements can't be avoided.

Other element quality checks include checks for warping out of plane, interior angles, mid-side node placement, and coincident elements.

There is a coincident node check to detect coincident nodes within a small tolerance supplied by the user. This command will optionally renumber adjacent elements so that they share the same nodes. This is called "merging" out the duplicate nodes. The program will ask if you also want to delete the unused nodes after renumbering the elements.

Bandwidth Minimization

For many finite element solvers, the solution time is significantly affected by the bandwidth of the model's stiffness matrix. If you look at the location of the numbers in a stiffness matrix, they tend to lie in a band about the diagonal of the matrix. Outside this band is all zeroes. The width of this band is called the "bandwidth," which is dependent on the way the nodes are numbered in the model. Since the solver does not have to store the numbers outside this band, the storage size is proportional to the bandwidth. The solution time will increase rapidly as the bandwidth increases.

The Model Solution task automatically performs bandwidth minimization internally (if required), but for other external solvers you will have to request this step. You can either generate a sequence list (which the solver will use internally), or you may permanently renumber the nodes of your model to minimize the bandwidth.

Some analysis programs use a different solution algorithm where the solution time is more proportional to the "wavefront," which is a function of the element numbering. Both the bandwidth and the wavefront can be checked or optimized.

It is recommended that you do node and element checking before you apply nodal boundary conditions, because renumbering or deleting nodes and elements often will change the definition of the boundary conditions.

Adaptive Meshing

Adaptive Meshing is a way to automatically change a mesh of nodes and elements to improve it. Often, a finer grid of elements is required in areas of high stress or strain energy. The Simulation application can refine a mesh using either analysis results or element distortion values as the basis for the modification.

Although there is a wide variety of different combinations of changes that are allowed, and datasets that could be used as the basis of the modification, it is recommended that strain energy be used for mesh refinement. Any analysis dataset can be used for the adaptive meshing, as long as it is in the form of scalar (not tensor or vector) element data (one value per element). Some datasets that come from the solution cannot be used directly if they are vector type data. You can create a new dataset under the Results, Create menu, taking one component such as the maximum value to reduce this dataset to a scalar value to use it for adaptive meshing.

The following example shows how adaptive meshing can be used to refine a mesh based on analysis results.

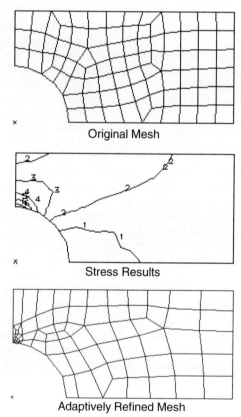

Original Mesh

Stress Results

Adaptively Refined Mesh

For more information on this topic, search for the word "Adaptive" in the Finite Element Modeling User's Guide in the Help Library.

Boundary Conditions Task

The Boundary Conditions task is used to build analysis cases containing loads and restraint boundary conditions to apply to the model. Boundary conditions can be applied to the part geometry before meshing or to the resulting nodes and elements. Applying boundary conditions to the part geometry will mean that if the part is changed and the model updated, the boundary conditions will also be updated.

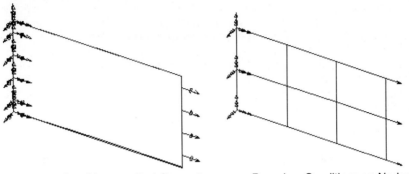

Boundary Conditions on Part Geometry   Boundary Conditions on Nodes

An analysis case is a collection of DOF sets, constraints, restraints, structural loads, and heat transfer loads. For most structural problems, you will only need structural loads and restraints.

Structural loads can be nodal forces (forces directly at a node) or pressures on the face or edge of an element (which are converted to nodal forces internally). A nodal force has six values, for the three forces and the three moments. To create nodal forces, you will first be prompted to select the nodes. To select multiple nodes, use shift-pick or select by a screen area.

Restraints are used to restrain the model to ground. Restraints also have six values at nodes: three translations and three rotations. Each entry can either have a value for the fixed displacement or is left free to move. Nodes are selected to apply restraints the same way as applying forces. The values given for restraints apply to the displacement coordinate system for the nodes, not the global coordinate system. A model should normally be held in space by restraints so that it is not free to move in any direction even if there are no applied forces in that direction, or the problem may not solve.

Forces and restraints are graphically illustrated using arrows on the model. Forces have open arrowheads and restraints have closed arrow heads. Rotations in either case have a double arrowhead.

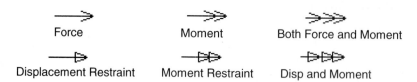

Force   Moment   Both Force and Moment

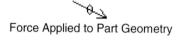

Displacement Restraint   Moment Restraint   Disp and Moment

Forces and restraints applied to part geometry are indicated by a circle around the arrow.

Force Applied to Part Geometry

Constraints sound like "restraints," but are not to be confused. Constraints are used to constrain nodes to other nodes, not to ground. They can be used to impose special cases of symmetry boundary conditions, or special relationships between nodes.

DOF Sets are used to identify special degrees of freedom in the problem, commonly used for dynamics. A set of degrees of freedom defined as a DOF set can be used in an analysis case as Master DOF, Kinematic DOF, or Connection DOF. Master DOF are used for one of the dynamic solution methods called Guyan Reduction. The problem will be reduced to this master set of degrees of freedom for the dynamic solution. Kinematic DOF can be used in static problems to reduce out overall kinematic freedoms in a model in the case where no boundary conditions exist in a particular direction, but it is generally better to use restraints as described above to fully restrain a model.

Common Sense on Boundary Conditions

Modeling errors (such as not enough elements or distorted elements) tend to go away if you follow the guidelines in iteratively refining the mesh until you have shown that you have converged on an answer. Boundary condition errors will not converge out no matter how much you refine the model. Also, boundary condition errors may not be obvious in the results. If you have elements that are not connected properly and a gap opens up in the model, a check of the displacements will show this; but the effect on the displacements due to the difference between a simple support and a fixed support is more subtle. For this reason, boundary condition errors are the most dangerous errors you can make. There is no substitute for careful checking. For example, you should always be able to do a simple hand calculation to make sure your results are not far from what is expected. Also, learn to look critically at the displacement plots for any unexpected changes in direction or slope of the deflection shapes. Any unexplained high stressed areas in the model may be due to a restraint placed at the wrong location.

There is a strong tendency for people to believe the results that come out of a computer without questioning the assumptions made for boundary conditions. The burden is on the Analyst to make sure that good assumptions have been made.

Symmetry and Anti-Symmetry

If a model is symmetric, and the loading is also symmetric, it is often possible to model only half of the problem. The trick is to apply the correct boundary conditions to force the displacements on the symmetry plane of the half model to be the same as the displacements that would have occurred in the whole model because of the symmetry conditions. Tremendous efficiency is gained by doing this, because not only is the model half the size, but also the bandwidth is often less as well, and the same problem can be solved in one quarter of the time or less.

If you can understand the discussion to follow, you will be one of very few that know how to properly use symmetry boundary conditions. If you can't, you should not try to use this shortcut, or you may get wrong results and not know they are wrong- a very dangerous combination.

At the plane of symmetry (if the structure and the load are symmetric), you could place a mirror, and would see the same thing in the mirror as what is behind the mirror (the same model, deflection, and stresses). What degrees of freedom don't move at nodes in the plane of the mirror? There can be no motion perpendicular to the plane, or rotations in the plane. An example of a case where symmetry could be used would be squeezing the prongs of a tuning fork.

A way to tabulate all the possible combinations of planes and degrees of freedom is to construct a table as follows with the degrees of freedom labeled across the top, and the rows labeled with the plane of symmetry. This discussion is clearer if the planes are identified by the normal to the plane. For example, in this chart the X=0 plane is what I-DEAS identifies as the YZ plane.

We will fill in each position in the chart with an "F" for free or a "0" for no motion, the same as the entry you will use to enter restraints in I-DEAS. First, fill in a "0" for the translations normal to each plane:

```
           Degrees of Freedom
  PLANE    X    Y    Z    RX   RY   RZ
   X=0     0
   Y=0          0
   Z=0               0
```

Next, fill in an "F" for the other translations since they are free to move:

```
           Degrees of Freedom
  PLANE    X    Y    Z    RX   RY   RZ
   X=0     0    F    F
   Y=0     F    0    F
   Z=0     F    F    0
```

The next step is a little harder to intuitively understand, but it is easy to see the pattern. Rotations in the plane cannot move. The pattern is to put the "complement" of the translations into the columns for the rotations (reverse F and 0), to complete the table for symmetric boundary conditions:

```
        SYMMETRIC BOUNDARY CONDITIONS

         Degrees of Freedom
  PLANE   X    Y    Z    RX   RY   RZ
  X=0     0    F    F    F    0    0
  Y=0     F    0    F    0    F    0
  Z=0     F    F    0    0    0    F
```

It is safer to learn to use this pattern rather than to try to figure out the correct boundary conditions by intuition each time for the particular plane of symmetry.

There is another case of symmetry where the model is symmetric, and loads have a symmetry, but the sign is reversed across the plane of the "mirror." This is called "anti-symmetric." This is NOT the same as "asymmetric," which means that there is no symmetry. Anti-symmetry occurs very often, and can again give you great savings in solution time if you can understand when and how to use it. An example of a case of anti-symmetry would be a twisting load applied to the frame of a car.

To figure out what boundary conditions to apply for anti-symmetry is even harder to understand intuitively, but the pattern is again very easy, and provides a safer method to make sure you get it right. All you have to do is take the complement of the entire table above:

```
            ANTI-SYMMETRIC
          BOUNDARY CONDITIONS

         Degrees of Freedom
  PLANE   X    Y    Z    RX   RY   RZ
  X=0     F    0    0    0    F    F
  Y=0     0    F    0    F    0    F
  Z=0     0    0    F    F    F    0
```

There are many cases where a problem will have more than one plane of symmetry. An example is a model to find the stress in a rectangular plate with a hole in the center under tension at the edges of the plate. This problem can be solved by modeling only one quarter of the plate. When modeling a problem where two planes of symmetry intersect at nodes on both planes, the boundary conditions on these common nodes should be restrained with a "0" if restrained by either symmetry plane.

Model Solution Task

The Model Solution task in the Simulation application is where the finite element model is solved. This chapter will introduce the standard types of analyses that can be performed, including statics, buckling, heat transfer, and potential flow. Other types of analyses include P-element, dynamics, and nonlinear analysis.

The Model Solution task is easy to use since no file transfers or additional steps are necessary (in contrast to the process of solving the model using other external solvers). The results are stored in the model file as "datasets" that can be displayed in the Post-Processing task. Unlike parts, the results datasets are not stored in libraries.

To use the Model Solution task, the finite element model must be created in the model file you are using as described in previous chapters. Boundary conditions and forces must also be created and selected into a boundary condition set. You should always save your model file before running the solution in case something drastic goes wrong during the solution, such as a power failure or a serious error.

 In the Model Solution task, you will first create a "Solution Set" which describes the type of solution, the type of output to store, and other options.

 Second, you may optionally set some solution options with the Manage Solve icon.

 Next, you will solve the model with the *Solve* command.

 After the solution is complete, you should review the errors and warnings.

After the solution successfully finishes, the requested datasets will automatically be stored in the scratch model file. If you want to permanently save them, either save your model file, or write a universal file. (You may not want to save until you have reviewed the results in the Post Processing task to see if they are correct. You also may not want to save at all, because it will make your model file larger. If you have saved your model file just before the solve, you can always solve the model again later if you need to get the results back.)

The Solution Set contains the type of solution (statics or dynamics, etc.), output selections, and other options. One option is the method of solution such as to request a "Verification Only" solution. If the problem is large, this will give you a useful estimate of the file sizes and solution time.

You may never need to set the solve options. In most cases, you can just keep the default settings and solve the model.

You have the choice of running the solution interactively (the default) or in batch mode. If you ask for a batch solution, a batch run is submitted when you give the *Solve* command, and you will be returned to the operating system. You can't access the model file until the solve is complete.

Another set of execution options pertains to the special scratch file which the Model Solution task uses to store and manipulate the matrices during the solution. This file is called the "Hypermatrix" file. This file is automatically created and deleted after the solution, unless you request that it be kept. Since this file can be very large, you will not want to keep it unless you need to run a restart analysis which allows you to supply new loads to a static run, or calculate more modes for a dynamic analysis.

If there is not enough room on the disk for the Hypermatrix file, the solution will abort. Model Solution does not put any limit on the size of the problem you can solve, but the practical limit is usually set by the amount of disk space you have available to hold the Hypermatrix file.

The Solve form also lets you control the name of the output list file (sometimes called a Log file). This file will contain a log of the solution steps, which will be helpful in diagnosing errors.

Another option in this form is a switch to tell the solver not to perform a bandwidth profile reduction before the solution. Normally, the program automatically does this. In special circumstances, if you want to deactivate this feature, a toggle button switch will turn this off.

To solve a finite element model for linear statics, you must have the model properly restrained. Improperly creating restraints is a common user error. Even if you don't apply a load in a particular direction, you must restrain the model against all six possible rigid body motions, or singularities will result and the solution will abort.

Singularities can also be caused by "cracks" in the model, where coincident nodes occur. These must be merged out, and the extra nodes deleted. It may look like the model is restrained, but one piece is floating off in space.

A more subtle cause of singularity errors is improperly connecting elements of different types. Properly connecting different element types requires a knowledge of the degrees of freedom used in the stiffness matrix of the element. For example, solid elements only use three degrees of freedom for the stiffness terms at each node, where beam elements use all six. Connecting a beam to a solid element at one node will result in a "ball joint," which may cause singularity errors, or worse yet, wrong answers that may appear correct.

Another relatively common problem that can cause the solution to abort is using the default thickness for thin shell elements, which may not be anything near what you want. If the other dimensions of the model are much smaller or larger, this may result in bending stiffness terms in the stiffness matrix that are so many orders of magnitude larger than other terms that numerical errors can occur.

Supported Elements

Like any other finite element solution program, Model Solution has a library of element types it understands. You must build your model using only these supported elements if you want to solve the model in Model Solution. A summary of the elements supported for the different analysis types are presented in the following table.

	Structural Analysis	Heat Transfer	Potential Flow
Model Solution Element Library			
Element Family			
Lumped Mass	X		
Springs	X		
Gaps[1]	X		
Rigid Bar	X		
Rod	X	X	
Beam (L,P)	X	X	
Plane Stress,Strain (L,P)	X		
Axisymmetric Thin Shell (L,P)	X	X	
Axisymmetric Solid (L,P, Q,T)	X	X	X
Thin Shell (L,P, Q,T)	X	X	X
Solid Elements(L,P, Q,T)	X	X	X
1 Gap Elements for static analysis only, not dynamics or buckling.			
Element order: L = Linear P = Parabolic.			
Element topologies: Q = Quadrilateral, T = Triangle.			

For a complete description of elements supported in the Model Solution task, see the manual "Simulation: Element Library Reference Manual" in the online Help Library. This manual gives important information on the internal degrees of freedom used in each element and describes how physical and material properties are used. For the more theoretically inclined, there are references to the formulation used in each element.

Post-Processing Task

The Post Processing task of the Simulation application provides tools to display and interpret the results after the solution is finished. Results can also be brought in from external finite element analysis programs for post-processing. Several different display types are available including contour plots and deformed geometry plots.

To make most displays, you will "walk" through the top row of three icons to select results, choose a display type, and set the element calculation domain; then select the icon Display.

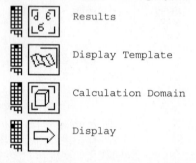

Results

Display Template

Calculation Domain

Display

Calculation results to be displayed are stored in "datasets" after the solution. Displacements will be stored in one dataset, and stresses in another. Depending on what you requested for output selections in the Solution Set, you may have other datasets stored as well, such as element forces, reaction forces, and strain energy. If you solve for multiple Solution Sets, you will have multiple sets of datasets from each solution.

You will use the *Results* command to select a dataset for display, and another dataset for deformations. The resulting display will show the display results superimposed on the deformed shape.

Different components of both stress and deformation can be displayed. Stress data is stored in a dataset as the raw tensor values, so that the software can calculate and display any component such as maximum principal stress, maximum shear stress, a specific directional component, or Von-Mises stress. The component of stress is selected in the *Results* command.

Visualizer

The *Visualizer* command starts a more advanced post processing task which takes advantage of hardware graphics to display output results. It is available only if you are using PEX or OGL (not X3D) when you start I-DEAS. The use of the Visualizer is presented in an online tutorial.

I-DEAS Visualizer

Display Template

The *Display Template* command is used to select the display type. The results display and deformed geometry can be turned on and off independently in this form. Results display types include contour plots, element criterion plots, and arrow plots.

Contour plots can be displayed as a line, hidden line, stepped shaded, or a smooth shaded display.

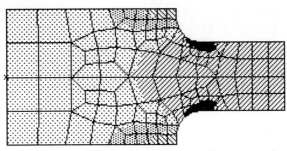

Element criterion plots will display elements that exceed some chosen criterion, such as displaying the elements with stresses over the yield stress, or the elements with stresses above 90% of the highest stress in the model.

Arrow plots display magnitude and direction as arrows on nodes or faces of each element. These plots can be very useful in getting an intuitive understanding of how stresses flow through the structure, and how this flow causes stress concentrations at certain locations.

Deformed geometry can be displayed with or without element borders, shown as solid or dashed lines. The undeformed geometry can also be turned on or off in this form.

Calculation Domain

The command *Calculation Domain* controls averaging across elements and which elements to use for the calculation. You can either select pre-defined groups of elements for the calculation, or leave the default set to the option *Selected Elements*. In this case, when you execute a display, the program will ask you to pick the elements to plot. Pressing Return or the middle mouse button will default to all elements being displayed.

It is important to understand that contour-plotting algorithms perform averaging on calculated values to produce a contour plot. You must be careful about relying on exact stress numbers read from a contour plot. For example, if two different elements have different materials, you may not want to average across this boundary. The *Calculation Domain* command will let you set where averaging occurs across element boundaries.

Display

After defining the display, execute the display with the command *Display*. This command will prompt you to pick elements for the display if the *Calculation Domain* option is left at the default of *Selected Elements*.

Often, you will want to change the *Display Filter* settings to make a clearer display. For example, you may want to turn off the wireframe geometry, the workplane, and the part display. The display of part surfaces and element surfaces can interfere with each other and produce a confusing or cluttered display.

Animation

Displays can also be animated using the *Animate* command. The frames of animation can be from one dataset, such as a static analysis or a mode shape, or from multiple datasets, such as different time steps from a transient analysis. The animation frames are first drawn to store them in the local terminal memory, then they are sequenced through to display the animation. To stop the animation, use the pop-up menu and select *End*.

XY Graph

Another form of display is to generate an XY graph of results. You can select specific nodes to plot a results dataset vs. the coordinate location or distance between nodes.

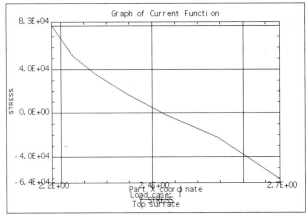

Deleting Results

If you want to change the model after looking at the results, you must first delete the analysis datasets for that model or copy the finite element model to a new model in the model file. The program will not let you change the model or boundary conditions on the model if datasets exist. This ensures that you can't possibly display results using the wrong model or boundary conditions.

Beam Sections Task

For many general structures, there are at least three completely different ways to make a finite element model. For the Offset Link created in Workshop 2B, three different finite element models are shown below. Solid elements could be used to make the first model shown. This model looks most like the real part, and little user interaction is required to generate it, but the solution time will be high due to the large number of degrees of freedom that must be solved.

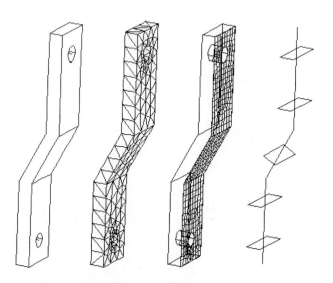

Since this part is relatively thin, thin-shell elements could be used to create the second finite element model. An assumption must be made by the user that the part is thin enough to model it using thin-shell elements. Using thin-shell elements assumes that the stress can only vary linearly through the thickness, but the resulting model will take much less time to solve since it has fewer degrees of freedom. The finite element model is geometrically modeled in two dimensions instead of three. The third dimension is entered as a physical property value for the elements, but is not displayed graphically.

Since the Link is relatively slender compared to its length, a further reduced model can be made using beam elements. These elements, (with only two nodes per element and six degrees of freedom at each node) are very general, but lead to smaller models in terms of the total number of degrees of freedom in the problem. The beam elements display one dimension (length). The other two dimensions are contained as properties of the beam elements.

Although beam elements are geometrically the simplest elements, they can be more difficult to use than solid or thin-shell elements. Some of the potential problems are: (1) the required properties can be difficult or tedious to calculate; (2) the cross section properties can easily be oriented the wrong way; (3) stresses (which vary over the cross section) can be difficult to interpret from tables of numbers; and (4) beams require more assumptions on the part of the user. Because of these potential problems, new users tend to favor other methods, but they are giving up the advantages of beam elements in doing this.

The Simulation application has some special tools to overcome most of the disadvantages mentioned above. Let's look at each of these problems in the order they were listed.

First, the required cross sectional properties are automatically calculated from the beam cross-section geometry. A unique feature of the software is that not only the resulting properties, but also the entire cross-section geometry is stored for the beam elements. Most standard beam sections are created from tables of beam shapes or by entering the key dimensions of different families of beam shapes. More general shapes are entered using a wireframe geometry representation of the boundary.

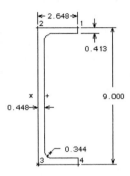

To make sure that the beam sections are oriented properly, the Simulation application can display a triad showing the coordinate system of the beams or display the actual cross-section geometry on the beam elements. A beam model might look like the figure on the left when displaying the beam elements as a line. The figure on the right shows the element triads turned on.

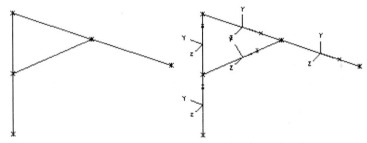

The same model is much easier to understand when the beam cross-section geometry is displayed, either as an outline or as a hidden-line-removed display. A Line display will show beams as on the left. A Hidden or Shaded display will show the beams as on the right. This type of display not only shows you if the beam is oriented properly, it gives graphical feedback that all its dimensions have been entered properly.

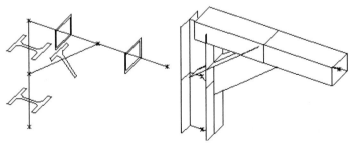

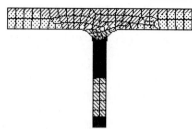

To help interpret the stresses on the beam sections, the Post Processing task can display different components of stress on the same cross-section geometry that was used to model the beam sections.

Judgment is required of the user, but the graphical display tools give the user intuition to know if using beam elements is a good choice, and where the model needs to be refined. The software does not take a part model and decide for you if using thin-shell or beam elements is a good assumption. Some of this judgment comes with experience, and some comes from having an inquisitive mind. In general, you should always question the effects of your assumptions. Rarely will one model be sufficient to answer all the questions that should be asked. The best course of action is to use a simple model first, and find out the problem areas, and then build refined models in these areas. This will also give you a relative feel for the accuracy of your models as you converge on an answer, which is much better than having only the results from one model.

All of the elements we have used in previous workshops use two sets of property tables; material and physical properties. Beam elements in the Simulation application have one more property, the cross section property, which stores the entire geometrical description of the cross section, not just the computed properties (like the area and moments of inertia). The cross section properties are defined in the Beam Sections task. The icon commands in this task are summarized at the end of this chapter, and demonstrated in the workshop to follow.

There is also some geometric data associated with each beam element such as end releases to model pin ends rather than welded, continuous ends. This beam geometric data is added to beam elements or modified with the icons:

One of the items stored with each beam is the method of orienting the beam section. Beam elements have a local coordinate system where X is aligned from the first node to the second node. The Y and Z axes are perpendicular to this. The default orientation is that the local Z axis lies in the plane formed by the beam local X axis and the global Z axis. In the case that the beam is aligned with the global Z axis, the beam's local Y axis is aligned with the global Y axis. If you don't want to use this default orientation, it is possible to specify a third node to orient the beam's local Y and Z axes using the *Orientation* command. You can also supply an orientation angle to rotate the beam coordinate system. The picture on the previous page shows some beam elements with the element triad turned on, showing the default orientation.

Cross Section Properties

For each different cross section, properties like the area and moments of inertia, Iy and Iz, are calculated from the section geometry. Using the Beam Sections task, all the properties required are automatically computed. For the "T" cross section shown, the following properties are calculated.

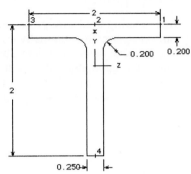

```
Section type          : Tee
Dimensions
 Depth                : 2.0
 Flange width         : 2.0
 Flange thickness     : 0.2
 Web thickness        : 0.25
 Fillet radius        : 0.2
Properties
 Area                 : 0.8672
 Prin. moment of inertia Y: 0.1360
 Prin. moment of inertia Z: 0.3399
 Shear ratio Y        : 2.7998
 Shear ratio Z        : 2.6015
 Torsional constant   : 0.0180
 Warping constant     : 0.0
 Warping restraint factor : 0.0
 Eccentricity Y       :-0.5218
 Eccentricity Z       : 0.0
 Plastic modulus Y    : 0.2310
 Plastic modulus Z    : 0.4831
 Plastic modulus torsion : 0.605
 Offset rotation angle : 0.0
 Rt                   : 0.5219
```

Stress recovery values:

Pt.	Code	Cy	Cz	Radius effective
1	2	0.6218	1.0	0.0
2	2	0.6218	0.0	0.45505
3	2	0.6218	-1.0	0.0
4	2	-1.3782	0.0	0.0

While beams are a very simple finite element type, the interpretation of beam stresses is not so simple. For other types of finite elements, stresses are computed and stored at the node points of the elements. For beam elements, the stresses vary over the cross section as well as along the beam. For this reason, there are two ways to output beam stresses. The stresses can be plotted on the beam cross section in the Post-Processing task, or you can display stresses along the beam at specific "stress recovery points". These points are labeled on the section plot with the numbers 1 through 4, above. Their Y and Z locations in the beam coordinate system are listed with the properties.

For more information about the stress recovery values, codes, and the actual formula used in each case, use the help option available when entering or modifying these values.

There are three types of beam cross sections: Standard Shapes, General Shapes, and Keyed-in data.

Standard Shapes include a library of standard shapes such as angles, boxes, channels, rectangular and round tubes, T sections, and wide flange beams.

You can either key in dimensions to describe these shapes, or select standard AISC (American Institute of Steel Construction) sections from a table. The cross section properties are computed from formulas for the specific cross section dimensions, so this is the fastest method for calculating the properties.

The *General* shape type is the cross section type you would use if none of the standard shapes applies. This section type allows you to describe the cross section from points, lines and arcs. The properties for general shapes are calculated using a temporary mesh of nodes and elements that is automatically created on the 2D section. The mesh will be stored as part of the cross section property table and used to display the stresses after the solution is performed.

When creating a general cross section, you can create the geometry manually from points and lines; you can select standard shapes to use to create this geometry; you can start with a standard shape and modify it; or you can combine standard shapes. The figure here is a sample of a general cross section with the mesh used to evaluate the properties.

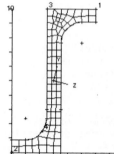

The last option for entering beam cross section data is to manually key-in the basic properties including the area, moments of inertia, torsional constants, etc. This option will obviously not allow you to plot stresses on the cross section later, because the program has no information about the boundary describing the cross section. You might have to use this method if you have only been given beam section properties, but don't have the geometric description.

Beam Display Options

To check the orientation of beam cross sections, the element triad or a beam section display can be turned on. These two display option switches can be turned on with the menu commands:

```
/DISPLAY_OPTIONS
    ELEMENT
        TRIAD SWITCH
            ON, OFF
    BEAM_SECTION
        ON, OFF
```

With the beam section switch turned on, beam sections are sketched at the center of the beam elements with the *Draw* command. If a hidden line or shaded image display is requested, a full representation of the beam geometry is displayed.

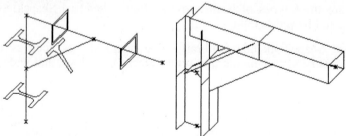

Solution and Analysis Datasets

Beam elements can return two different types of information, both related to stresses and both stored as analysis datasets available for use in post processing. These are "element forces" and "element stresses." If you use the Model Solution task to solve the model, you will generally want to request that both of these datasets be stored, along with displacements.

Element forces are stored as an analysis dataset containing beam forces and moments at each end of the beam. You must store element forces if you want to plot stresses on the cross section of a beam element.

For other types of elements, you would normally want to store element stresses. If your model contains only beam elements, however, you may not want to store this dataset, since these stresses are only tabulated values of stress components evaluated at the specific stress recovery points on the cross section. Beam element stress datasets are useful for scanning locations of maximum stress, but element force datasets must be used to plot detailed stresses across the beam cross section.

Beam Post Processing

Normally you should first check the deflections of the structure before trying to interpret stresses. Errors in boundary conditions or element connections are usually easier to detect in a deflection plot than in stress plots.

After checking that the overall deflections look reasonable, you would look at stresses in the overall model to locate the beams with the highest stresses. A stress contour plot will show the lines displaying the beam elements in the same colors shown in the color bar to represent values of stress. Remember that these stress values are determined at one particular stress recovery point. Be careful how you interpret these numbers. It is important to recognize that the stress contour algorithm averages the stress values (depending on what data component you request) at the nodes where the elements connect.

To get more complete and accurate stress values on an element, plot the stresses on the cross section using the command *Beam, Contour on X Section*. Select a beam element and the stress component to display stresses on a cross section. To use this capability, select the analysis dataset containing the element forces as the display dataset.

Shear stresses and torsional shear stresses are often best understood on an arrow plot, which indicates the magnitude and direction of the shear stresses.

The Post Processing task can also plot force and stress diagrams similar to traditional shear and moment diagrams used to analytically calculate beam deflections and stresses. A group of elements can be selected to plot these values on an XY plot format.

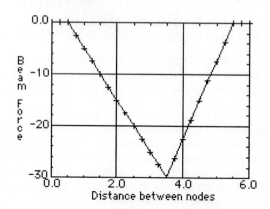

This same plot can be placed along the elements of the finite element model, to give a better intuitive feel for the correspondence of the plot and the physical structure using the commands in the menu *Beam, Force & Stress*.

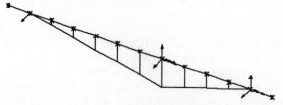

The Post Processing task also has the capability of performing standard code checking to automatically process beam forces against handbook equations for the maximum allowable loads.

Line stress distributions can also be plotted along the beam outline. The figure below shows how the shear stress varies along the Y axis of the T section.

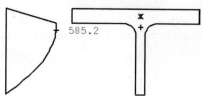

Optimization Task

Engineering design is an iterative process. You will normally want to change your design based on the results of the analysis. If there are many possible variables that can be changed, such as thickness, radii, and other part dimensions, this can be a lengthy trial-and-error process. The Optimization task can help in this process by performing parameter studies to tell you which variables will have the most effect and by putting Model Solution "inside a loop" to iteratively refine the model automatically.

Since the Optimization task can use the original design variables on the part as optimization parameters, you will want to construct your part using Variational Geometry and Relations as described in earlier chapters. Any dimensional constraints used to construct the part can be used as a variable parameter for optimization. As any variable is changed, all of the design intent rules that you designed into the part definition will be used to update the part geometry between iterations.

The optimization can be geometry-based or FE-based. Geometry-based optimization allows you to select part dimensions as optimization design parameters. FE-based optimization allows you to optimize finite physical properties, material properties and beam sections. For a geometry-based optimization, you must turn on the "geometry-based analysis only" switch when you create the FE model. This will ensure that only geometry boundary conditions (as opposed to nodal boundary conditions) are applied so that loads will be automatically updated with part geometry changes.

You should run at least one normal solution before using the Optimization task, to make sure your model and boundary conditions are correct.

Users are tempted to immediately put the optimization solver to work with lots of design parameters and let it run many iterations until convergence. A more intelligent use of shape optimization is to use the parameter study option or to run one iteration to perform sensitivity analysis on each design parameter to determine which parameters have the greatest effect, and to use the solver as a tool to give you insight into the design analysis.

Optimization is very efficient, due to the fact that the gradients of each design variable are computed internally by the solver. It only takes the solver one iteration to predict the effect of making a design change, and a few iterations to converge on an optimal solution.

The graphs below are samples of the history of an optimization solution of a problem with three design parameters. The first graph shows the changes to the values of the three parameters. The second graph shows the history of the stress.

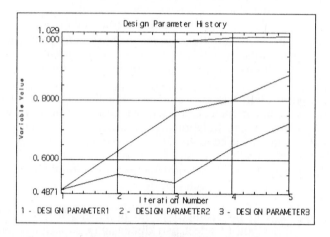

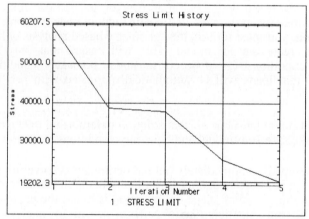

Laminates Analysis

Laminates analysis involves both pre- and post-processing. The laminate properties are defined from ply material information and the stacking properties in the Laminate task. Preliminary loads analysis may be performed on the laminate prior to a finite element analysis to determine if the laminate has the desired characteristics by plotting failure envelopes and stress/strain profiles through the thickness direction of the laminate. When the final laminate lay-up is defined, the resulting material properties are used in a shell element FE model. The results of this model can be interrogated in the Post Processing task to plot ply-by-ply stresses, failure indices, and strains.

LAMINATES ANALYSIS STEPS

Define Material Properties

Create material property tables to represent the matrix and fiber properties that will make up the plies.

Define Ply Properties

Create the ply material property tables using the basic material properties defined above or effective orthotropic properties of the ply. The program will also ask for other properties such as matrix and fiber volume fractions and ply thickness. The program will create a material property table and a ply property table to represent a ply. The ply material table will include allowable stresses if loads analysis is to be performed and failure is to be evaluated.

Create a Laminate Definition

Ply groups are defined, specifying orientation, thickness, and ply property table for each ply in the group. Different methods of defining the stacking within a ply group are available such as Symmetric, Antisymmetric, repeating plys. Ply groups are stacked using the same options as plies within a group. This results in a laminate definition table. The laminate can be displayed along with the constitutive matrix terms that define the relationships between loads and laminate strains and curvatures.

Loads Analysis

Loads analysis is an optional step to plot failure envelopes, ply-by-ply. This can be useful for predicting the first ply failure and determining the optimal lay-up for the desired structural characteristics.

Assign Laminate Properties to FE Model

Modify the shell elements to use the laminate material table created above. Material orientations must be defined for each element using one of several methods such as keying in the angles, vector projection, or by translating or revolving a pre-defined curve using a method called "dynamic curve tangent."

Perform the Finite Element Analysis

Make sure to store Stress Resultant.

Interrogate Results

In the Post Processing task, display model deflections. Create a dataset containing ply stresses from the strain dataset computed by Model Solution. Create a dataset containing ply failure index from the ply stress dataset. These datasets can be plotted ply-by-ply.

For more information on laminates, search in the Finite Element Modeling User's Guide for the word "laminate."

Dynamics

Dynamics (or vibrations) in a system are the result of energy being transferred back and forth between kinetic energy (energy stored in the motion of masses) and potential energy (energy stored in spring stiffnesses). A dynamic system with springs and masses would vibrate forever if it were not for energy being dissipated (lost) in damping, usually modeled as viscous damping, where the damping is proportional to velocity. The equations of motion for a dynamic system can be written in matrix form:

$$\{F\} = [K]\{X\} + [C]\{\dot{X}\} + [M]\{\ddot{X}\}$$

where $\{F\}$ is a vector of forces on each degree of freedom (DOF) in the system, $[M]$ is the mass matrix, $[C]$ is the viscous damping matrix, $[K]$ is the stiffness matrix, $\{X\}$, $\{\dot{X}\}$, and $\{\ddot{X}\}$ are the displacements, velocities, and accelerations of each DOF. This equation can be solved by directly integrating it at different steps in time, or by first solving for the modes of vibration characterized by natural frequencies and mode shapes, and using these to calculate the dynamic responses. The I-DEAS software uses this second approach to solve the dynamics of systems.

When the modes of vibration are solved, the resulting equations of motion are similar:

$$\{F\} = [\backslash K \backslash]\{x\} + [\backslash C \backslash]\{\dot{x}\} + [\backslash M \backslash]\{\ddot{x}\}$$

but now the mass, damping, and stiffness matrices are diagonal, making the mathematics much simpler, which is why this form is often used. The degree of freedom vectors $\{x\}$, $\{\dot{x}\}$, and $\{\ddot{x}\}$ above now do not represent physical DOF of the system, but "modal degrees of freedom". If we know the values of the modal DOF, the physical DOF can be calculated by:

$$\{X\} = [U]\{x\}$$

where each column of the matrix $[U]$ is a mode shape corresponding to each modal degree of freedom. These two equations together are called the "modal form" of the equations of motion. This matrix $[U]$ can be called a constraint matrix, since it constrains the physical DOF $\{X\}$ to the independent modal DOF. This makes them dependent. The modal degrees of freedom are the independent DOF in the equation. In this modal form, the number of independent degrees of freedom has been reduced from the number of physical DOF in the model (which may be thousands) to the number of modes (where less than one hundred may be enough even for a very complicated model).

Dynamic Analysis in I-DEAS

To solve a dynamics problem in I-DEAS, after creating the model you will (1) create boundary conditions, (2) solve for the modes, and (3) calculate and display responses. In each step, you must keep in mind what you're going to do in following steps. For example, if you only want to compute mode shapes and frequencies, you may set up boundary conditions and solve the model differently than if you want to compute a response to a base motion.

Creating Boundary Conditions for Dynamics

In the Boundary Conditions task, you need to consider if you intend to compute responses, or only compute mode shapes and frequencies. Also, you need to know if you intend to use the standard mode displacement method or the more advanced mode acceleration method.

Define the analysis type as either *Normal Mode Dynamics* or *Response Dynamics*. *Response Dynamics* is the most general. Select *Normal Mode Dynamics* if you know you only want to compute mode shapes or will only compute responses due to forces, not due to ground motion. Select *Response Dynamics* if you will use ground motion or if you will use the mode acceleration method to compute responses.

Tip: Do not select *Forced Response*. This only applies for a problem being defined for an external solver.

If you intend to compute the response to ground motion such as a seismic analysis, you need to define a set of connection degrees of freedom in the directions of the input motion. A special set of constraint modes will be calculated for these degrees of freedom to allow the base motion in these directions. Restrain other degrees of freedom that don't move.

If you intend to use the mode acceleration method of computing responses, you also need to create a set of kinematic degrees of freedom at the points where you intend to apply forces. These will be used to compute attachment modes, which will account for residual flexibility in the higher frequency modes not calculated.

When you create a boundary condition set, select either *Response Dynamics* or one of the normal mode methods to be described later. The boundary condition types that don't apply will be grayed out. For example, you can't select any load sets, because normal mode dynamics does not depend on loads.

Solving the Modes

Dynamics problems are solved to compute the natural frequencies and mode shapes in the Model Solution task. This creates the modal form of the equations of motion as shown previously. These mode shapes can be plotted as a deformed geometry display in the Post Processing task or using the Visualizer.

When you create a solution set, select one of the three methods for solving for natural frequencies and mode shapes (normal modes) only, or select *Response Dynamics*. The three methods are SVI (Simultaneous Vector Iteration), Guyan Reduction, and the Lanczos method. Response Dynamics also uses the Lanczos method, but also calculates constraint modes and attachment modes.

To solve for modes only, use the Lanczos solver as your first choice. This solver has several advantages over the Guyan or SVI methods. It requires less user input to set up the solution set, and is generally faster than the SVI method, especially for large problems.

To solve for normal modes of vibration using the Lanczos method, select *Normal Mode Dynamics - Lanczos* as the type of solution in your solution set. Select *Solution Options*, *Solution Control* to enter specific controls, such as the number of flexible modes and the frequency accuracy desired. In the *Output Selection*, make sure mode shapes are to be stored.

Guyan Reduction is a classical method of solving dynamics by reducing the problem to a set of "master DOF." This reduced set of matrices is solved, and then the mode shapes are expanded to the other degrees of freedom. Guyan Reduction requires the user to select a set of "master degrees of freedom" into the boundary condition set in the Boundary Conditions task. The mass and stiffness matrices will be reduced to these degrees of freedom before solving the dynamics problem. Since a reduced problem is being solved, this solution algorithm can be very economical, especially if a moderate number of well-chosen masters is selected. The accuracy of the results, however, will be dependent on how well you distribute these degrees of freedom on the model.

The SVI method uses a Kinematic Degrees of Freedom (KDOF) set in the boundary condition case set to solve for rigid body modes. Enough kinematic degrees of freedom must be included to restrain the rigid body modes if these locations were used as restraints. Selecting three translational degrees of freedom at three non-collinear nodes will suffice to control the rigid body modes. The number of rigid body modes expected must also be entered in solution set options.

The accuracy of the SVI method does not depend on any user judgment, but the solution will normally take longer than Guyan Reduction except in the case where too many master degrees of freedom are selected for the Guyan Reduction.

Response Analysis

Next, the equations of motion are solved to calculate time or frequency responses to given inputs in the Response Analysis task. You can enter excitation functions and define response points to evaluate and graph motions and stresses as XY functions of time or frequency. Motion can be plotted as displacement, velocity, or acceleration functions. Stress can be plotted as any component of the stress tensor at a node point.

To use the Response Analysis task, you will normally first create functions to be used as input forces or motions.

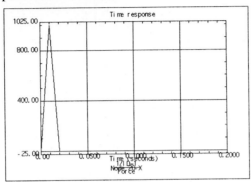

Typical Force Input Function

You will next create an "event" which defines the type of response you want to calculate. This can typically be *Transient, Frequency, Random*, or *Response Spectrum*.

The event will also define the modes being used for the calculation, and the functions to be used for input. Since modal damping is not calculated, you will manually define the damping for each mode when you define the event.

To compute the response at nodes, use the icon *Nodal Functions*. When you press *Generate Functions*, you will be prompted to pick the nodes where you would like to see the responses calculated.

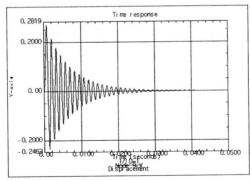

Typical Time Response Function

Another type of function that can be evaluated is Frequency Response Functions (FRF) as shown below. An FRF is the ratio of output/input at given DOFs graphed vs. frequency. To calculate an FRF, create a frequency event and include a function with a value of 1.0 for the entire frequency range. FRF are usually graphed as magnitude and phase on a logarithmic scale as shown below.

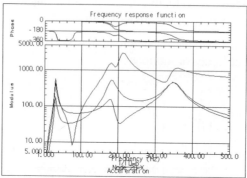

Typical Frequency Response Function

Evaluated functions can be further manipulated with math operations such as integration, differentiation, and FFT using the menus under the *Function Tools* icons. Time and frequency functions can also be exported for use elsewhere, such as to perform other processing on them in the Test application.

The Response Analysis task can also compute detailed stresses and deflections of the model at "snapshots" in time, using the icon *Evaluate Results*. The resulting datasets can be displayed in the Post Processing task or the Visualizer. The prompt will ask you for the specific time or frequency to evaluate, and for a number of intervals up to this value.

Other Solution Methods

Previous discussion in this chapter described linear statics and dynamics. There are a variety of other solution methods also available. These are available as solution options in the Manage Solve form in the Model Solution task.

Durability Task

The Dynamics task estimates fatigue life based on static or dynamic events, as defined in the previous section on dynamics. Results can be calculated as the number of event cycles to failure, and can be plotted as contour plots.

P-Method Linear Static Solution

In normal finite element analysis, you converge on a solution by increasing the number of elements. The P-Method is a different element formulation that converges on a solution by increasing the order of polynomial equations within the element. (The "P" comes from polynomial.) The advantage of this method is that (1) a fewer number of larger elements can be used, and (2) the convergence is an automatic part of the solution, with less user interaction.

In the Model Solution task, linear statics problems can be solved by either the conventional approach or by using this P-Element method.

In using the P-Method, you must be careful not to include interior corners that physically would lead to infinite stress concentrations. In conventional finite element modeling, users tend to get sloppy and leave out small radii if they are not expected to be a problem. While the stress at these locations is not correct, the deflections in the overall model, and stresses elsewhere in the model, still can be accurate. With the P-Element method, however, the solution will attempt to converge on the infinite value.

For more information on this method, search for the word "P-Method" in the Model Solution/Optimization User's Guide.

Nonlinear Analysis

One of the solution types in the Model Solution task when you create a Solution Set is "Nonlinear Statics." This option allows you to solve problems with nonlinear material properties and geometric nonlinearities. Examples of nonlinear material properties are a nonlinear stress-strain curve and creep. Geometric nonlinearities can be caused by displacements large enough that the stiffness matrix needs to be reformulated at several intervals.

For more information on this topic, search for the word "nonlinear" in the Model Solution/Optimization User's Guide.

Buckling Analysis

Buckling analysis will compute a requested number of buckling mode shapes, and calculate how close the given load is to the critical buckling load. This analysis does not predict what will happen after buckling starts, only when buckling instability will be reached. The results for each buckling mode shape are expressed as a load factor, which is the amount by which the given loads in the load case must be multiplied for buckling to occur. If this load factor is 2, this means that the given loads only bring the structure to 50% of the critical buckling load.

Heat Transfer

The Heat Transfer solution type solves for steady-state thermal analysis. Temperatures, flux, and reaction heat sources are stored as datasets for display in Post Processing. Load Sets can also be generated to use as loading conditions for a linear statics structural analysis.

When the model is created, a case set must be created containing element loads of flux, convection, or heat generation, and nodal heat sources. Restraints are created as nodal temperature restraints.

Heat transfer in the Model Solution task only performs linear conduction problems. For non-linear radiation problems, I-DEAS provides an interface to TMG from Maya Technologies. For information, see the online manual "Simulation: TMG Thermal Analysis User's Guide."

Another specific task in the Simulation application is Electro-System Cooling. This task analyzes convective cooling of electrical packages.

Potential Flow

Potential flow analysis is used for predicting the flow of a fluid through a region. The flow of the fluid is characterized by a scalar field termed the velocity potential. If the velocity potential is known, other important data can be calculated such as velocity and pressure.

As with other types of analyses, a case set must be defined before executing the solve. Velocity potential restraints are applied to the model as X direction restraints. Velocity flux is entered as element face pressures or edge pressures. Source terms are applied as nodal forces. Since structural items are used for the entry of these quantities, use only SI units to prevent units conversion from incorrectly scaling to these flow quantities.

Plastics Analysis

The Simulation application also includes the tasks Thermoplastic Molding, Thermoset Molding, and Weld Locator to analyze the injection molding process to design good parts and fill systems and to optimize injection molding process parameters to give minimum cycle times.

Interfacing Other FE Codes

To solve your model in an external solver such as ANSYS or NASTRAN, you must *Export* an "input deck" for this program. (For those of you who don't know what a "deck" is, this terminology is a carryover from the old days when input was prepared on a deck of punched cards.)

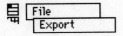

To solve your model in a program not included, you could write an interface using a universal file, IGES, or in one of the other supported formats. Universal file formats are documented in the online manuals.

After the solve, import the results file back into I-DEAS. The *Import* command will give you a menu of analysis programs to read from. The results will be converted to universal file format, which you can read into I-DEAS.

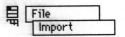

If you are running your analysis on a different computer, you will need to use a stand-alone version of the Dataloader program to convert the output file into a universal file, which can be transferred to the computer where the I-DEAS software is running.

For more detailed information on file translation or the Dataloader, see information in the online documentation in the manual "Simulation: External Solvers User's Guide."

Summary

This chapter has covered a lot of material on finite element analysis using the Simulation application. It has given a broad background on the tools available for geometry creation, meshing, boundary conditions, solution, and post processing. This chapter mainly focused on linear statics analysis, but other analysis types such as dynamics were also discussed. Although there was an introduction to some other capabilities such as optimization, P element analysis, buckling, laminates, and adaptive meshing, you will need to consult the online manuals for more details on these topics.

Where To Go For More Information

For more information on the information presented in this chapter, see the following information in the Help Library Bookshelf:

Help, Help Library
    Simulation: Finite Element Modeling User's Guide
        General Simulation Topics
            Simulation Overview
            Simulation Tools
        Boundary Conditions
        Meshing
        Applying Boundary Conditions
        Post Processing
        Visualizer
        Beam Sections
        Laminates
    Model Solution/Optimization User's Guide
    Element Library
    Model Solution Verification Manual
    External Solver User's Guide

The Simulation online tutorials cover a broad range of topics from fundamental skills like meshing and boundary conditions through more advanced topics like thermal and buckling analysis.

Help, Help Library, Tutorials
    Simulation
        Fundamentals
        Advanced Projects

Also, remember you can search for the terms mentioned in each section of this chapter, in the appropriate user guide. Use *Help, On Context* to find command descriptions of the commands mentioned.

Help, On Context

The following is a sample of topics not covered in this chapter. To find information on these topics, use search or look in the Index or Contents of each book.

Data Surfaces, Data Edges
Time and Temperature Variations
Applying Accelerations
Coupled Degrees of Freedom
Section Meshing
Contact

⚡ PARTING_SUGGESTIONS

Geometry Modeling

-Before creating a model, consider: what type of analysis will be done? What results are expected? What questions are to be answered by the model?

-If the geometry came from a detailed part model, has it been simplified down to what is important? (Suppress small features in the part that need not be included in the FE model.)

Meshing

-Avoid distorted elements and quick transitions from small elements to large elements next to each other.

-Large aspect ratio elements (ratio of long side divided by short side) should be avoided. They are worse when the angles are other than ninety degrees.

-Use smaller elements where higher stress gradients are expected.

-Have elements been chosen that can adequately represent the expected response of the model? Would a 2D model suffice? What type of element is required? Beams? Thin Shells? Solid? Will your solver support the type of element you have chosen?

-Have different types of elements been properly connected?

-Use groups to help organize the model and to make it easier for others to understand.

-Don't be fooled into thinking that mapped mesh is better than free mesh because the grid looks more regular. The distortion is often actually higher.

-Don't make the model more detailed than necessary. It would be better to gain insight from several simple models than to spend all your time making one complex model.

⚘ PARTING_SUGGESTIONS

Boundary Conditions

-Remember - errors in modeling such as element size or type will tend to converge out as you refine the model. Errors in boundary conditions will not. If you have the wrong boundary conditions, you may converge on an answer by refining your model, but you will be converging on the wrong answer.

-Use your model to question the significance of the assumptions made about loading, boundary conditions, etc.

-Have different types of elements been properly connected?

-Are your boundary conditions properly modeled? Are the assumptions good? Can symmetry be used to cut the model size? Has the model been properly restrained in all directions?

-How well do you understand the failure mechanisms? What failure criteria should be considered in your model: Yield? Buckling? Deflection? Fatigue?

Solution

-Remember to select the output datasets to store for later processing. For example, you normally should store displacements and stresses. It is a good habit to also store reaction forces. These can be checked to make sure they equal the applied forces. Store element forces for beam elements.

-Sources of singularities (could not solve): Model not properly constrained. Coincident nodes causing "cracks" in model. Invalid physical or material properties. (Using the default thickness for thin shell.) Different element types not properly connected (Beams to Solids). Note: Gap elements do not remove singularities. Incorrect material properties can also cause singularities.

-Other solution problems: Not enough disk space for Hypermatrix file.

-The Model Solution task cannot use all elements that the Meshing task can create. Be sure to check that elements to be used are supported in the element library before creating the model.

-With I-DEAS 8 or later, you may need to edit the user_param file and insert the following line if the solve does not work. This will allow you to manually change the *Memory Usage* parameters, if needed.

```
Memory.AutoSetting: 0
```

PARTING_SUGGESTIONS

Post Processing, Interpreting Results

-Does the model answer the original questions?

-In checking the results, display deflections first. It is usually easier to see modeling errors in deflection plots than in stress contour plots. Can you compare maximum deflections to handbook formulas? Do the deflections look reasonable (no sharp discontinuities)? Do the slopes at boundary conditions look as expected? Were the boundary restraints properly applied?

-Interpretation of Stress Contour plots: Can you double-check the results by comparing with stress handbook formulas? Do the stresses look reasonable (no unexplained changes of contour angles that follow grid lines)? Are there enough elements at high stress locations to give accurate results? Are there any unexpected high stress areas in the middle of the model? (This could indicate wrong boundary conditions or forces.)

-What stress component most closely matches the failure criterion you should use?

-Are the results valid? Has the yield stress been exceeded anywhere in the model? Has buckling been included? Should large deflection theory be used?

-Be careful about reading exact answers off a contour plot, since the plot algorithm may do some averaging. For exact numbers, make a report of raw element values.

-Contour Color Bars: If you are using the X3D device type driver, turn off double buffering in Preferences, Display to get more colors for the contours. Double buffering gives smoother animation and dynamic viewing at the expense of fewer available colors.

ᵠ PARTING_SUGGESTIONS

Beam Elements

-Learn to use beam elements to make efficient finite element models, rather than taking a brute-force approach and making a detailed solid model, especially for the first analysis. The tools available make this possible to accomplish.

-Remember to request that the element force dataset be stored during the solution if you have beam elements in your model.

-A good way to interrogate beam section stress plots is to use two viewports; one to display the overall structure and one to plot the section views. This will eliminate the time required to redraw the structure to select the beam sections to plot.

-If your model contains beams and other element types, make sure they are properly connected. Connecting a beam element at one node to a solid element (which only has three DOF per node) will effectively give a ball joint.

Dynamics

-To use Model Response, you must first solve for the modes (Natural Frequencies and Mode Shapes).

-Remember to store stresses when you solve for the modes if you will want to compute dynamic stresses.

-Add damping values to the model before evaluating responses.

-Beam stresses cannot be graphed as responses, since their stresses are not stored in the same stress tensor form as other elements.

Other Solution Options

-Use linear analysis to check your model, boundary conditions, materials, and other assumptions before using the advanced options discussed in this chapter such as optimization and non-linear analysis.

-Check optimization sensitivities before attempting to blindly pick a large number of optimization design parameters.

-Solve very simple models at first with only a few elements to make sure you understand all the inputs and options for advanced methods.

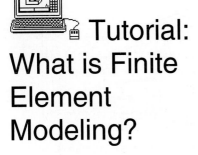

Tutorial: What is Finite Element Modeling?

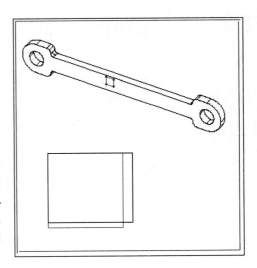

Although the concept of finite element modeling is to subdivide a complex model into a "finite" number of elements, it is informative to investigate in detail what happens to one element.

Before you start:

1. You should know the basics of I-DEAS, learned from earlier tutorials.

What you should learn:

1. The process of creating a finite element model, applying boundary conditions, solving, and displaying results.
2. The basics of how the program solves for displacements at each node and then computes stresses in each element.

Tutorial Instructions.

Start I-DEAS with a new model file and follow the instructions in the tutorial below.

Help, Help Library, Tutorials
Simulation
Simulation Projects
What is Finite Element Modeling?

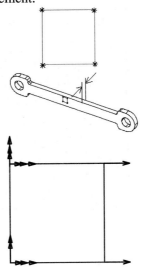

In this tutorial, you will create one element representing a piece of material in a linkage. You will apply boundary conditions and solve for the displacements and stresses.

Will the results be accurate with only one element? (Try it and see.)

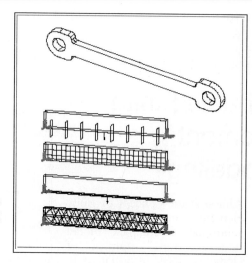

Tutorial: Which Element Type Should I Use?

Part of the "art" of finite element modeling is knowing which element type to use.

Before you start:

1. You should know how to model simple parts. If you need more modeling practice, see the tutorials in the list:

    Simulation
    Modeling Fundamentals

What you should learn:

1. How to use solid, thin-shell, and beam elements.
2. The advantages and disadvantages of each method.

Tutorial Instructions.

Start I-DEAS with a new model file and open the tutorial below.

Simulation
Simulation Projects,
Which Element Type Should I Use?

In this tutorial, you will apply a bending load to a linkage using several models with different element types. How do the displacements of each model compare to a hand formula? The stresses? (Don't forget that the hand formulas are based on some assumptions!)

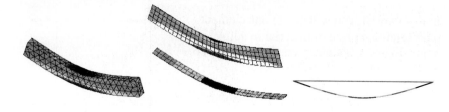

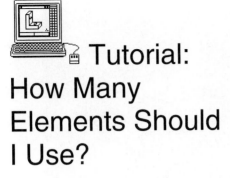

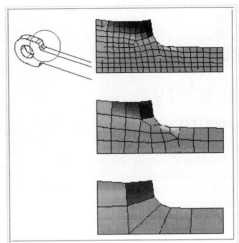

Tutorial: How Many Elements Should I Use?

Another "art" of finite element modeling is knowing how many elements to use.

Before you start:

1. You should know how to model simple parts.

What you should learn:

1. How to plot the results to estimate convergence, and recognize singularities, which will not converge.
2. How to estimate the accuracy of a model.

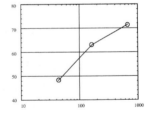

Tutorial Instructions.

Start I-DEAS with a new model file and open the tutorial below.

Simulation
    Simulation Projects,
        How Many Elements Should I Use?

By modeling the stress in the linkage at a stress concentration with different numbers of elements, you will see how the accuracy of the results is dependent on the number of elements used. How many is enough? Well it depends...

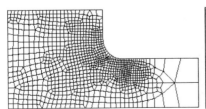

 Tutorial:
Boundary Conditions

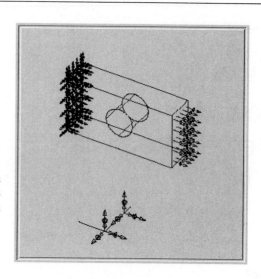

The largest source of errors in finite element modeling is probably improper application of boundary conditions.

Before you start:

1. You should know how to create and solve simple finite element models.

What you should learn:

1. How to properly apply boundary condition loads and restraints.
2. How to use symmetry to reduce the size of models.

Tutorial Instructions.

Start I-DEAS with a new model file and follow the instructions in the series of tutorials below to learn good techniques of applying boundary conditions.

Simulation
    Simulation Projects,
        Boundary Condition Sets
        Boundary Condition Surface Loads
        Boundary Condition Symmetry

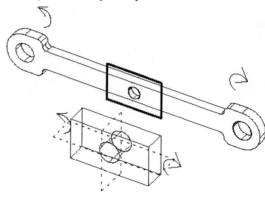

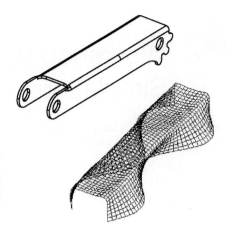

Tutorial: What's the Problem?

Finite element analysis is most commonly used for linear static analysis. Many other failure modes are common, and should be analyzed.

Before you start:

1. You should first complete the basic Simulation tutorials on meshing, boundary conditions, and displaying results.

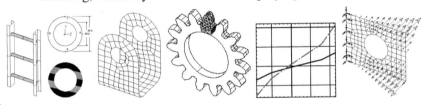

What you should learn:

1. How to use other analysis methods and analyze other failure conditions such as buckling, thermal analysis, dynamic response, and fatigue durability.

Tutorial Instructions.

Try the tutorials below for more advanced analysis methods.

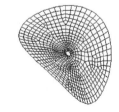

Simulation
    Simulation Projects,
        Optimization Parameter Studies
        Optimization Redesign
        Linear Buckling
        Thermal Analysis
        Contact Analysis
        Static Durability Life estimates
        Response Analysis
        Dynamic Durability Life Estimates

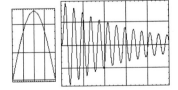

Workshop 9:
Simulation

This workshop will let you practice the techniques described in the earlier tutorials using the geometry of the pump parts created in earlier chapters.

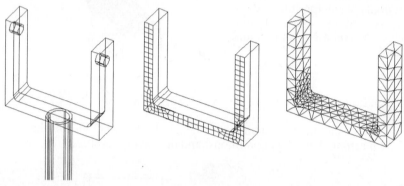

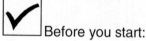Before you start:

1. Complete the earlier Simulation tutorials to learn the finite element modeling process and modeling techniques.

2. You should have the pump parts created in Workshop 2A. If you don't have these parts in a file or library, go back to Workshop 2A and create the part "Pivot Support."

What you should learn:

1. Review how to create different types of finite element meshes on your part geometry.

2. Review how to suppress unwanted detail from a part model.

3. Review how to solve the model and display results.

 The instructions in this workshop are not as detailed as the online tutorials described in the previous few pages. Pay attention to the prompt and list windows, and apply what you learned in the tutorials.

 Workshop Instructions:

Start the Simulation application, the Modeler task. Set your units to mm.

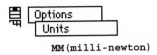

MM(milli-newton)

If you have the parts in a library, get a copy of the Pivot Support from the library. (Do not reference the part, since you will make changes to it.)

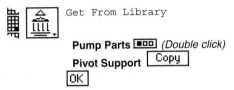

Pump Parts ▣□□ *(Double click)*
Pivot Support Copy
OK

Get the Pivot Support out on the workbench, and zoom in on the upper end as shown.

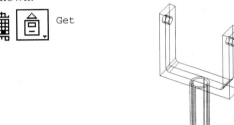

Suppress the pipe extrusion at the bottom. We will make a finite element model of just the upper yoke section of the part.

Modify
Click twice on side_ of tube.
Accept ▢■▢
Suppress Feature

Update

Suppress the extrusion that was used to cut the holes. We will assume that we don't need this detail in the finite element model we will make.

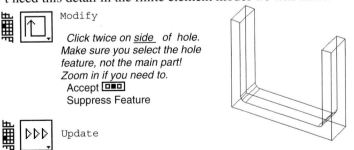

Modify
Click twice on side_ of hole.
Make sure you select the hole
feature, not the main part!
Zoom in if you need to.
Accept ▢■▢
Suppress Feature

Update

Create a finite element model named FEM1 associated with this part.

 Meshing

 Create

Part or Assembly: **Pivot Support**

FE Model Name FEM1
OK

Set the mesh parameters on the front surface, and mesh this surface, using thin-shell elements.

 Define Shell Mesh

Select front surface.
Element Length **10**

 Shell Mesh

Select front surface.

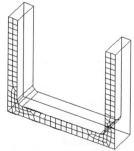

Make another finite element model associated with the same part, named FEM2.

 Create

FE Model Name FEM2
OK

Mesh the volume of the part, this time using solid elements.

 Define Solid Mesh

Select part volume.
Element Length **25**

 Solid Mesh

Select part volume.

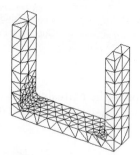

Look in the Manage Bins form. Notice that both FEM1 and FEM2 are stored with the part.

 Manage Bins

Pivot Support...
(Double-click to open and close.)

Put this part away.

The next model will be a thin-shell model of half the part. Rather than cut the existing part with a plane cut, you will sketch the geometry of the section.

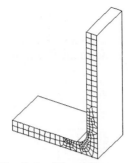

Sketch a shape as shown to represent one half of the Pivot Support as modeled in Workshop 2A. (Use the odometer to make the width about 4 inches, the height 6 inches, and the thickness about .5 inches.)

Modeler

Options
Units

inch(pound f)

Polyline

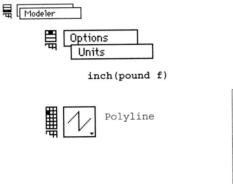

Fillet the internal radius to .5 inches.

 Fillet

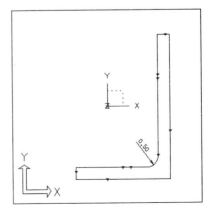

Delete and create dimensions as required to look like the figure shown below. Modify the values as shown. (You may use the *Appearance* command to display dimensions to two decimal places.)

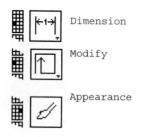 Dimension

Modify

Appearance

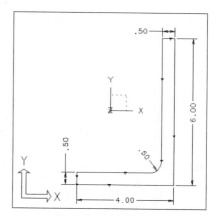

Extrude this shape 2 inches to make a part. (This is a convenient method to generate surfaces for 2D meshing. An alternate method is to use the command *Surface by Boundary* or another command to create a surface inside the region.)

 Extrude

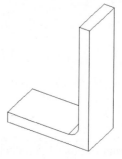

Name this part "Pivot Section."

 Name Parts

Create a finite element model associated with the part "Pivot Section" to generate the same model using free meshing. Turn on "Geomery Based Analysis," since you will apply geometry-based boundary conditions.

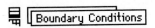

 Create FE Model

■ Geometry Based Analysis Only

Fully restrain the left edge.

 Displacement
Restraint

Pick entities   *(Pick left vertical edge shown below.)*
Pick entities   (Done) ▢■▢
    (Leave All Displacement Amplitudes Zero.)
    OK

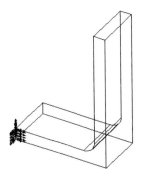

Apply a force of 1000 pounds in the X direction at the top left corner.

 Force

Pick entities *(Pick Vertex in upper left corner, below)*
Pick entities   (Done) ▢■▢
    X Force **1000**
    OK

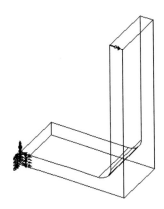

Define the front face to be a free mesh with an element length of .25 inches, and a 5 percent deviation at curved edges. Generate the mesh. (Your mesh may be slightly different than the figure shown.)

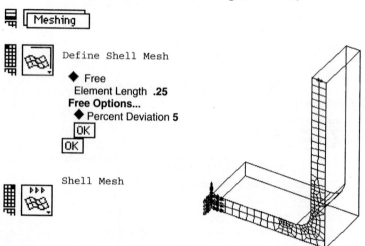

```
    Meshing

    Define Shell Mesh
        ◆ Free
        Element Length  .25
        Free Options...
            ◆ Percent Deviation 5
            OK
        OK

    Shell Mesh
```

To finish this model and prepare it for solution, modify the Physical Property table defining the thickness of the thin-shell elements. Define the thickness to be 2 inches. (The Physical Property Table "Thin Shell 1" was automatically created when you created the elements.)

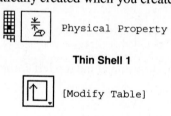

```
    Physical Property

    Thin Shell 1

    [Modify Table]

    Thickness    2
```

Modify the fillet radius to be .25 inch. Select the *Update* command to update the part and the model. Notice that the finite element model updates with the part.

Modeler

Modify

(Pick the part.)
(Accept)
Show Dimensions

Modify

(Pick the radius dimension.)
.25

Update

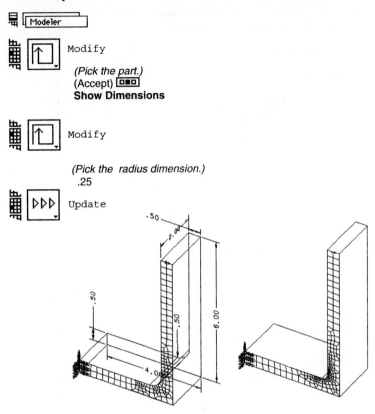

Create a solution set, which defines the type of solution to be performed. It also contains the boundary condition set and output selections. If more than one solution set has been created, this form will be used to select which one is "current." Select that displacements, stresses, reaction forces, and strain energy be stored after the solution.

After you have created a Solution Set, you can examine it or modify it. Pull down on the Type of Solution to see what other solution types are available. Set the solution method to "Verification Only". This will give some useful estimates for solution time and the size of the files required.

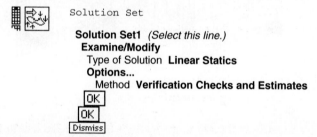

Look in the Manage Solve form to see what the options are. In this form you can define where your solution files will be placed. This is often used if you must place the "Hypermatrix" temporary solution file on a scratch disk instead of in your own directory. Selecting a verification run, above, will give an estimate of how large this file will be.

Solve the model (which in this case is only a verification run).

 Solve

During the verification, messages to this list window will be noted as "warnings." This will include a summation of all the element masses. If the material density is set wrong, or the element thickness is wrong, this number will reflect the errors. Since this value is mass, not weight, multiply by the acceleration of gravity, which for inch units is 386 to get the weight.

List the solution warnings, which in this case are only informational. An estimate for the solution time and the maximum Hypermatrix file size is included. Note that the time estimate is not the total time estimate, but only the decomposition phase, which for large problems is the major portion of time.

 Report Errors/Warnings

Change the solution method back to "Solution No Restart" in the Solution Set form.

 Solution Set

Solution Set1 *(Select this line.)*
   Examine/Modify
      Options...
         Method **Solution No Restart**
            OK
         OK
      Dismiss

Solve the model, this time to compute values.

 Solve

After the solve, check for solution errors.

 Report Errors/Warnings

If there are errors, you will have to go back through the workshop to find the problem. The most common errors are improper restraints or wrong physical or material properties.

Optional:

If you use other finite element programs such as ANSYS or NASTRAN, export the model so that you can perform the solve in that program.

 File
Export

After the solve, import the results file back into the Simulation application.

 File
Import

Display the default results using all elements.

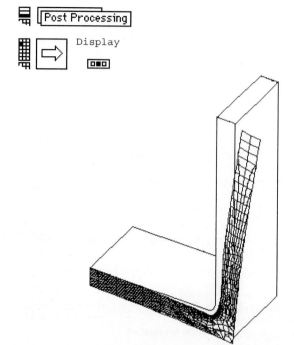

Depending on your display device, if you only see a limited number of colors, turn off double buffering in Options, Preferences, Display. This will make dynamic viewing and animation less smooth, but will allow more colors for the display.

Turn off the display of the part to make the displays less cluttered. Also turn off boundary conditions *(But remember to turn them back on later if you try to display any part models in the model file!)*

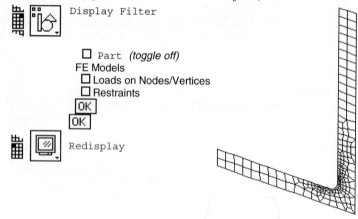

Display the Reaction Forces as an arrow display.

Results

Display Results

Reaction Forces ▷

(Select this line above, then press ▷)
OK

Front View

Display Template

◆ Arrow

Display

Pick Elements

Select and display the stresses.

Results

Display Results

Stress ▷

(Select this line above, then press ▷)
OK

Display Template

◆ Contour
Stepped Shaded
OK

Display

Pick Elements

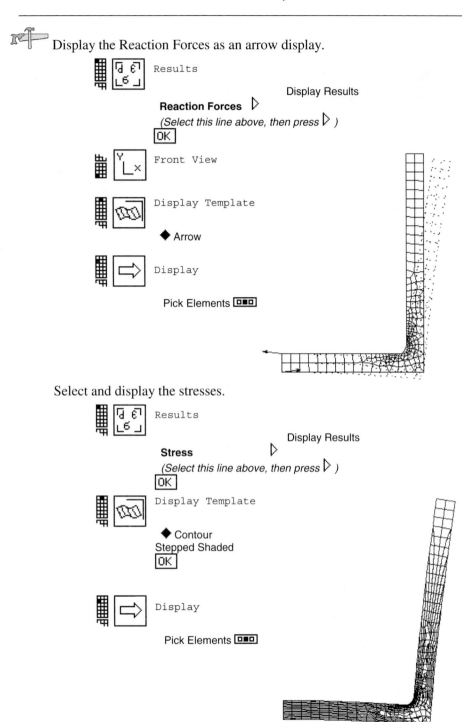

Select the Maximum Principal component of stress for the display.

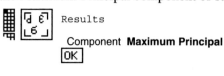 Results

Component **Maximum Principal**

OK

 Display

Pick Elements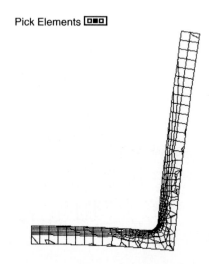

Zoom in on the corner to look at the stresses around the fillet.

  Zoom

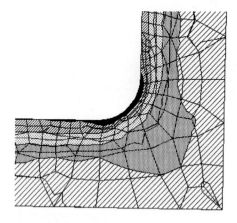

Use the *Probe* command to interrogate stress values at specific locations. (You must turn on the *Probe* switch in the *Display Template*.)

Display Template
    ■ Probe

Display
    Pick Elements ▢▣▢

Probe

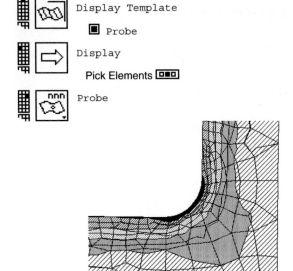

Display the maximum principal stress as an arrow display.

Display Template
    ◆ Arrow *(Turn on)*
    Arrow...
    (Press the Arrow Button for options.)
      Location ◆ Face
        OK
     OK

Display
    Pick Elements ▢▣▢

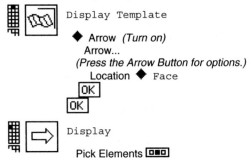

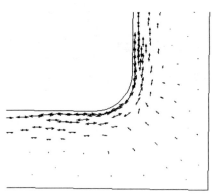

Display the elements with a stress over 40,000 psi.

 `Display Template`

◆ Element *(Turn on)*
Element...
(Press the Element Button for options.)
◆ `Results Units`
Above **40000**

OK

 `Display`
Pick Elements ▢▢▢

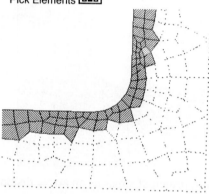

Display the finite element model geometry.

  `Line`

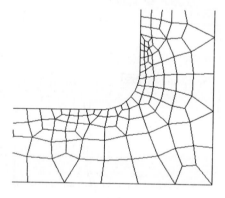

Make an XY graph of the Y stress component plotted against the X location through the thickness of the model along a line of nodes. (Your mesh may be different.)

  Setup XY Graph *(Pull down to get this icon.)*

Select Results to Plot
  Component **Y**
  OK
  Single Result Set
Pick Nodes/Elements
(Use Shift-Pick to pick nodes shown below)
Pick Nodes/Elements (Done) ▣▪▫
  Set Axis Type
    Node Data
      Part Coordinate
        X Direction
  Execute Graph
  Done

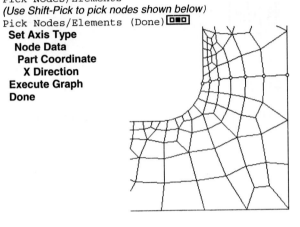

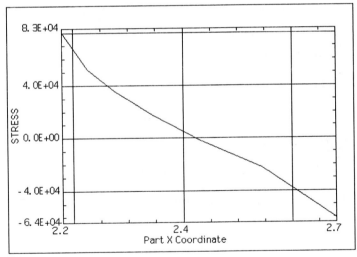

Turn part and boundary conditions display switches back on.

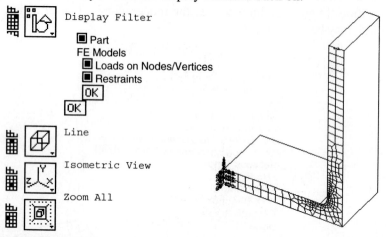

Display Filter
■ Part
FE Models
    ■ Loads on Nodes/Vertices
    ■ Restraints
    OK
OK

Line

Isometric View

Zoom All

Switch to the Optimization task to perform a parameter study on the effect of the fillet radius.

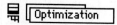 Optimization

Create an optimization "Design."

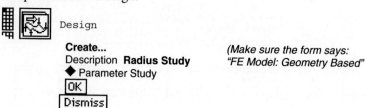

Design

Create...
Description **Radius Study**   (Make sure the form says:
◆ Parameter Study   "FE Model: Geometry Based"
OK
Dismiss

Create a Solution Set to describe the Load Case(s) for the design. Accept the default output selections.

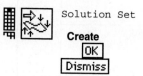

Solution Set
Create
OK
Dismiss

Define the Step Control to perform two positive and two negative iterations from the present value.

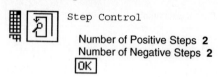

Step Control

Number of Positive Steps **2**
Number of Negative Steps **2**
OK

Define a Design Parameter for the fillet radius.

   Design Parameters

Create
Design Parameter Type **Geometry**
Description **Radius**
Select Dimension
Pick Dimensions Show Dimensions
Pick Dimensions *(Pick radius dimension)*
Step Size **.1**
OK
Dismiss

Define a stress monitor to graph the maximum stress in any element.

  Stress Monitor

Create
Description **Max Stress**
Select Geometry
  (Pick meshed surface.)
OK
Dismiss

Define a displacement monitor to graph the X deflection at the top.

  Displacement Monitor

Create
Description **X Deflection**
Displacement Direction **X**
Select Geometry
  (Pick vertex at top.)
OK
Dismiss

Define a mass monitor to graph the mass of all the elements.

  Mass Monitor

Create
Description **Mass**
Select Geometry
  (Pick front surface.)
OK
Dismiss

Save your model file and solve. During the solve, the part will change on the screen.

  File
Save

  Solve

(The solve may take several minutes...)

Note: With I-DEAS 8 or higher, you may need to edit the user_param file and insert the following line if the solve does not work. This will allow you to manually change the *Memory Usage* parameters, if needed.

```
Memory.AutoSetting: 0
```

Check for errors. Don't worry about informational warnings which are status messages during the solve.

Report Errors/Warnings

Use the *History* command to graph the stress, displacement and mass monitors as a function of the design parameters.

History

Select all three Monitors.

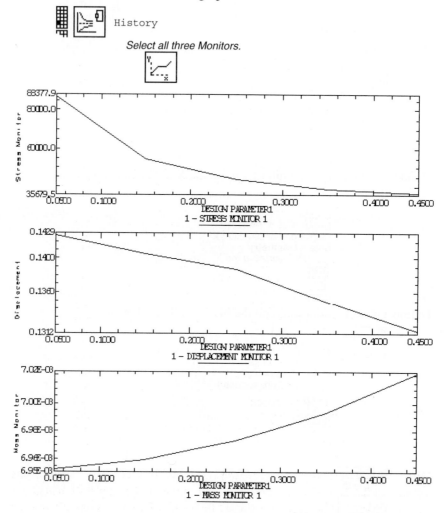

What effect does the fillet radius have on stress, deflection, and mass?

The graphs show that the radius change only makes about a 1% change in mass, a 7% change in displacement, but a 100% change in the stress.

Exit from the software when you are finished. This model file is not needed in any other tutorials.

Chapter 10
Other I-DEAS Applications

This chapter introduces some of the more highly specialized applications in the I-DEAS software suite. Not all product designs will use these software applications, but you should be aware of what is available in case you ever do need them.

These topics are not covered in the same depth as in earlier chapters. Tutorials and workshops are not included in this Student Guide, but training is available through online materials and classes.

Harness Design Task

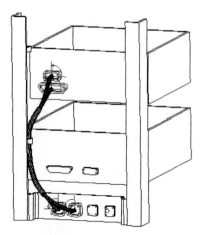

A wire harness designer must accurately model wire and cable bundles, connectors, power boxes, and other electrical and electronic modules to reserve adequate space for the harness within the product.

I-DEAS Harness Design is one of the tasks in the Design application. It can help you determine the space requirements and wire lengths for a wiring harness. You can use Harness Design to avoid interference between the wiring harness and other components in the product assembly.

The harness model is modeled in the context of an assembly, as covered in Chapter 6. A harness is defined by physical components and usually a separate logical wire list. The harness will add specific components such as connectors, paths, and bundles to the assembly, which already contains instances of the physical product. The same rules of assemblies apply, such as constraining connectors in the assembly.

Electrical engineers are typically responsible for the "logical" or electrical characteristics of the harness. These include connectivity, wire and cable attributes, and the overall electrical analysis of the system. A wire list is usually generated using separate electronic design software. This list will contain descriptions of the logical end points of wires and wire diameters.

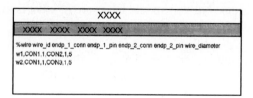

Once you design your harness, you can create several types of manufacturing drawings. For example, as well as assembly drawings showing the physical routing of wire bundles, you can display the wire harness straightened to a flat display.

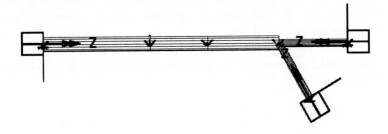

Where To Go For More Information

You should learn the basics of assembly modeling before Harness Design. After you learn to use the Assembly task, you will have a good background to read and understand the online material about this task.

Help, Help Library
  Design User's Guide
    Harness Design

Also, use *Help, On Context* to directly look up any command in the Command Descriptions.

There is also an online tutorial available.

Help, Help Library, Tutorials
  Design – Assemblies, Fundamentals
    Creating a Basic Wire Harness

This tutorial will demonstrate the basic concepts of creating a harness using simple parts.

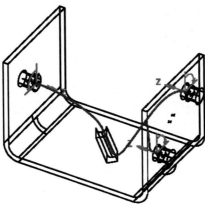

Sheet Metal

The sheet metal commands in the Modeler task can design sheet metal parts and fold and unfold them. The unfolded part can be used with the Generative Machining task of the Manufacturing application to generate toolpaths to make the part, and the Drafting task to document the flat pattern of the part.

The sheet metal commands are contained in a sub-icon panel in the Modeler task.

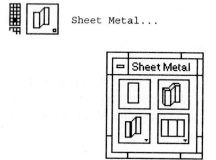

Sheet Metal...

Sheet Metal Panels

The sheet metal model is defined by panels connected by bends (with a radius) or welds.

Panels can either be defined from wireframe geometry or from the surfaces of a solid part. For example, the Offset Link part shown below was created in Workshop 2B by extruding the section shown on the left and then cutting holes. The sheet metal part on the right was created using the *Sheet Metal* icon, and picking three faces. The sheet metal part is thickened using the *Shell* command.

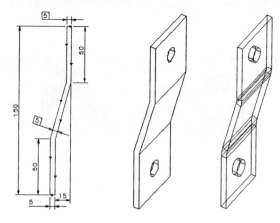

Another method to create a sheet metal model is to pick wireframe geometry to define panels using the *Build Panel* icon. This command can work two ways. You can either pick closed wireframe sections to create panels, or pick an open section, which will be extruded and thickened into a sheet metal part. The program will display arrows on the screen and prompt you for the extrusion and thickening directions. For example, the polyline curves shown below were picked with this command to create the sheet metal part shown on the right.

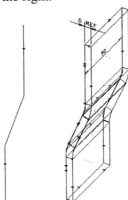

Sheet Metal Bends

Panels are joined by bends or welds. You can either create the bends when you create the panels, or create them later. After they are created, you can modify the radius values and K factor with the *Modify* command.

One attribute of a bend is a stress relief cut-out to avoid tearing at the ends of the bend. Several different forms of stress relief shape are available.

The K factor of a bend defines where the neutral axis will lie between the inner and outer radii. This factor means that the folding and unfolding operation takes into account the stretch of the material. This factor is normally about (.25). For a discussion of this topic, search for the topic "Controlling Bend Allowance" in the online help system.

Welds are like bends, only the radius is zero. Enter a zero radius to create a weld. (Older versions of the software had a separate *Weld* command.)

Sheet Metal Tool Catalog

Sheet Metal Design also includes a customizable tool catalog to create form features. You can create your own features to create things like holes, louvers, imprints, form features, add-on components, and inserts.

Folding and Unfolding

One panel should be grounded, so that when the part is unfolded, this panel will remain fixed in space. By default, the first panel created will be grounded. Sheet metal parts can be folded and unfolded by modifying the sheetmetal node in the history tree. Modification options include *Model Fold* and *Model Unfold*.

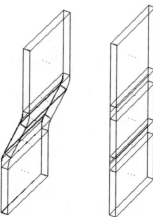

Creating a Solid

The sheet metal part is a special wireframe display. To create a solid part, use the *Shell* command. This can be done from the folded or unfolded state to create a solid part that can be used for Drafting, Assembly, Simulation, Generative Machining, or any other application where you would normally use a part model. For example, you might want to machine the part in the unfolded state, but display it in an assembly in the folded state.

Where To Go For More Information

For more information on sheet metal, see the following in the Help Library.

Help, Help Library
    Design User's Guide
        Sheet Metal Design
            Modeling Sheet Metal Parts
            Design Considerations
            Modifying Sheet Metal Parts
            Sheet Metal Tool Catalog

Also, use *Help, On Context* to directly look up any command in the Command Descriptions.

Mold Design

Look around you and you will find many parts that have been molded or cast out of plastic or other materials. For every one of these parts, someone had to design not only the part, but also the molds to make the part.

The design of the mold can be more complicated than the part itself. In fact, to keep the part simple, it may be worth the cost to use a complicated mold assembly. For instance if snaps can be molded into the parts instead of using screws for assembly, it may save a lot in the overall cost of the product.

The design of molds involves calculating shrinkage, defining the parting lines, checking draft angles, adding runners and gates, designing cooling systems and ejection systems, and designing the entire assembly that makes the mold. The mold may have side-actions that move as the mold opens. These may need simple mechanisms to create this motion.

The key to the process for creating molds in I-DEAS is to create a partitioning tool to split a mold into mold halves and side actions.

As well as the general modeling tools, I-DEAS contains two specific products- VGX Core Cavity and VGX Mold Base for mold design.

VGX Core Cavity

The VGX Core Cavity commands provide tools for checking draft angles for moldability, creating the parting surfaces, and partitioning the mold.

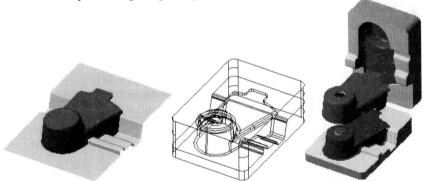

The VGX Core Cavity *Die Lock Check* command is a very useful diagnostic tool. By coloring the faces for the mold halves, molding problems are much easier to find visually. These colors are used by the *Parting Surface* command to identify sides of the mold. The *Mold Maker's Partition* command splits the mold.

VGX Mold Base

The VGX Mold Base commands provide specific tools for creating a mold base for injection plastic molding and adding mold components to it.

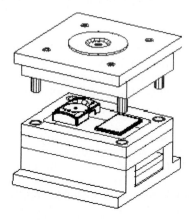

This software is much more than a catalog of mold plates and components. This software lets you build a mold assembly around your parts to be molded on the workbench. As you add components such as cooling lines and ejection pins, you always work in a 3D model to avoid interferences. VGX assembly constraints are used to position components so that the parts stay properly positioned as you change dimensions.

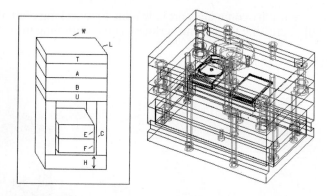

As you enter dimensions on a form, the mold is previewed in 3D around your parts. You can easily see if the mold is big enough to contain the parts and to leave room for cooling lines and other components.

Mold Filling Analysis

To analyze the filling and cooling times, there are two Moldflow products available from within I-DEAS. Moldflow Part Advisor uses a geometric model of the part to do a quick analysis of the filling of thin-walled parts. This software does not require you to build a finite element mesh.

Moldflow Plastics Insight performs a more rigorous analysis using finite difference and finite element analysis. This analysis requires you to build a finite element model as covered in Chapter 9. Specific element types are added to model cooling lines and runners.

Where To Go For More Information

For more information on mold design see the online manuals.

> *Help, Help Library*
> > *Design*
> > > *Mold Design User's Guide*
> > > > *Mold Design*
> > > > > *VGX Mold Design Software*
> > > > > *VGX Core Cavity Software*

Also, use *Help*, *On Context* to directly look up any command in the Command Descriptions.

Test Application

In a CAE approach to a design project, the goal is to do as much simulation of product performance in the computer as possible rather than performing expensive, time-consuming tests. There still is an important role for testing, however. By integrating the test results into the same computer database, and making these results accessible to a design team, more effective use of test results can be made throughout the design process. Initial testing may be done to understand loads on the part due to the environment it will see. Testing may be done later to verify a finite element model.

Test Application Tasks	
Time History	Perform Time History and Function creation and manipulation. Perform math operations and graphing. Includes Statistical operations.
Histogram	Reduce Functions to frequency distributions (Histograms).
Model Preparation	Create Node and Element or Trace line geometry for mode shape displays. (Similar to Meshing task in Simulation.)
Signal Processing	Perform Data Acquisition, Perform Random and Sine data processing, perform Order Tracking and Harmonic Tracking Analysis.
Modal	Extract mode shapes and damping from measured FRF (Frequency Response Functions).
Structural Mod.	Modify the modal model, predict sensitivities.
Correlation	Compare experimentally measured and analytically predicted mode shapes.
Fatigue Life	Estimate life to crack-initiation using measured strain data.
Post Processing	Display mode shapes (similar to Post Processing task in Simulation application).

Functions

Much as the Design application creates and displays parts, and the Simulation application creates and displays finite element models containing nodes and elements, the basic entity that the Test application manipulates is functions. A function is a list of related (X,Y) data pairs, such as a displacement on a structure measured as a function of time.

The Test application can create functions or read them from data acquisition equipment. It can perform math and statistical analysis on these functions and plot them in different forms. As well as providing a general toolkit of data processing capability for functions, the Test application also has specific analysis capability in the areas of modal testing of structural dynamics, rotating machinery analysis, fatigue life estimation from strain data, and advanced spectral analysis.

Function Attributes

When a function is stored, function attributes are stored with it that describe the attributes that pertain to the abscissa and ordinate quantities, whether the function is evenly or unevenly spaced, and how many data points are contained, etc. These attributes can be listed or modified if appropriate. At times, it may be important to know the function attributes, since incompatible attributes may prevent some math operations. For example, to add two functions together, the attributes must be the same.

The X axis of a function, called the abscissa, can be time, frequency, RPM, distance, or any other quantity to label the horizontal axis of the plot. The abscissa values in a function are either evenly spaced or unevenly spaced. If the values are evenly spaced, the individual values are not stored by the program. Instead, the starting value and the increment are stored with the function.

The Y axis (ordinate) of a function is the contents of the function to be plotted or manipulated. The units of the Y axis can be any quantity. Common measured quantities are displacement, velocity, acceleration, or strain. The contents of a function can be real values or complex. For measured quantities, they are real. If we take the Fourier transform of a function to convert it from the time domain to the frequency domain, the contents at each data value will be a complex number, containing magnitude and phase.

Management of Functions

Functions are managed slightly differently than other entities in the I-DEAS software. Since tests may include very large quantities of test data, it would not make sense to store all this data in the model file where everything else is stored. A new type of file called an "Attached Data File" (ADF) is used to store these functions. The name of this file will be stored in your model file, and when you re-open your model file, the ADF file last used will automatically be opened.

Temporary Array scratch space is used to store, plot, and manipulate functions in the model file, but the ADF is intended for permanent storage of functions. The functions stored in the function Array space in the model file are accessed by a number and a name.

Function Operations

The *Function Operations* menu commands and corresponding icons in the center of the icon panel allow you to create functions, edit functions, perform single-function and multiple-function math operations, and perform statistical analysis.

Function Creation Methods Create

Although the intention of Test is to analyze measured data, functions can be created in the function task by several methods. These methods include:

Function Creation Methods	
Keyboard Entry	Enter data points manually.
Random Number	Fill the function with random numbers.
Digitize From Grid	Use cursor on a grid to enter function.
Digitize From Data	Use cursor to pick points from an existing function.
Algebraic	Key in an expression.

In all methods, a form will prompt you to set the function attributes first. These will be the attributes given to the new function.

In using the keyboard entry method to create a function, you can type ! to back up to previous entries if you realize you made a mistake after pressing Return.

In using algebraic entry, you can use your own defined variables (described in Appendix B) or special variables to indicate the X axis (Z_X) or other stored functions (Z_Y1 for function 1, Z_Y2 for function 2, etc.).

Function Math Operations Single Math Multi-Math

Single-function math operations are operations that work with one function. Examples of single-function math operations are Trigonometric functions, Differentiation, Integration, Linear Interpolation, and Scaling. Linear Interpolation is used when you want to perform multi-function operations between two functions that do not have exactly the same abscissa spacing. In all, there are over fifty different single-function math operations available.

Multi-function math operations include simple operations like adding, subtracting, multiplying, and dividing two functions.

Function Editing Edit

The *Edit* command lets you delete or replace a range of a function or concatenate multiple functions together.

Function Statistics

A general toolkit of statistics analysis tools is provided in the *Individual Statistics* icon and the stack of icons under it. Analysis tools exist to calculate statistical deviation properties of one function, to calculate a correlation coefficient between two functions, or to perform polynomial regression analysis, to fit a polynomial through data points. A polynomial of any order from 1 to 20 (1 is the same as linear regression) can be fit to the requested function. The coefficients of the polynomial are listed, and the resulting function is evaluated and stored.

XY Graphing

XY graphing is a core capability in the Test application. The XY Graphing icons are found at the center of the icon panel. The default graph format will be automatically chosen based on the type of data being graphed. If you want to change the graph style, you can either set Grid Options and Data Options directly or choose a graph format from the icon command *XY Gallery*.

XYZ Graphing

Multiple functions can be graphed as a three dimensional surface. The Z axis will be either the function number or the Z Axis Value, if supplied as one of the function attributes. Surfaces can be plotted as line, hidden line, or shaded image.

A concept of function management useful for 3D plots is the concept of a "Record Trace." A Record Trace is a list of records on the function ADF that are grouped together. These records can be manipulated, graphed or deleted by the name of the Record Trace. When you ask for an XYZ Graph, you can plot a range of functions or select the functions by a Record Trace.

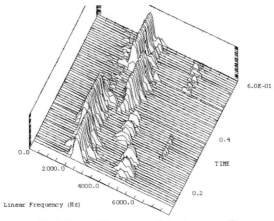

Bird Song - Frequency Spectrum vs. Time

Time History Task

The Time History task contains commands for interactively editing long time history records. Time History records are used for strain recordings or raw time history data to be used to calculate Frequency Response Functions (FRF). One of the operations that can be performed on a time history record is digital filtering. An example of using digital filtering to remove low frequency recording noise from a recording of a bird song is shown below. This figure shows one time segment of the data used to calculate the spectrum vs. time graph shown on the previous page.

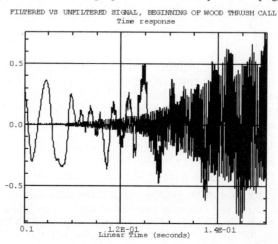

Histogram Task

A histogram is a way of plotting the amplitude distribution of a function by dividing the Y axis into "bins" and counting the number of events that fall into each bin. Different methods are available to determine what counts as an event. The simplest method is to count each data point. Other methods are available such as looking at level crossings or the time in each bin. Methods of counting amplitude cycles between high and low values, used in fatigue life estimation, are also available.

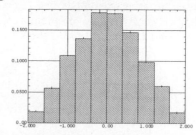

Statistics is a quantitative way of looking at the same distribution information as shown in histograms. Statistics and histograms are closely related. Statistics can be calculated from a function, or from a previously calculated histogram. A histogram can also be calculated from the statistical information, to plot the statistical information in a graphical histogram form.

Modal Analysis Task

Modal Analysis is a way to characterize the dynamics of a structure as a combination of "modes" of vibration, each of which can be described by the parameters of natural frequency, damping, and mode shape. There are two ways to get these parameters. One is analytical, such as using the Simulation application. The other is by performing a test on an existing structure. The analytical method is subject to modeling errors. The major limitation of the test method is that a structure must exist for testing. The best approach is to use both, if possible, and use the test to check the accuracy of the model. This is why the Test application includes the capability of both deriving the modes from test measurements, and tools for comparing test and analytical results.

The basic measurement used for a modal test is the Frequency Response Function (FRF), sometimes called a Transfer Function. This function represents the ratio of the response of a point on the structure (usually acceleration) divided by the input at another point (usually force) plotted as a function of frequency. This is a complex function, containing magnitude and phase. The magnitude of this function will have peaks at the natural frequencies of the structure. The height of each peak relative to other measured FRFs is proportional to the mode shape. The width of each peak is related to the damping. The higher the damping for a particular mode, the wider the peak will be.

Two important function attributes of FRF data taken for a modal test are the "Reference" and "Response" coordinates. The Reference coordinate is the node number and direction of the point that is constant in each FRF measurement. If a shaker is used to excite the structure from one fixed location, the shaker attachment point is the Reference coordinate, and the Response coordinates are where accelerometer measurements are made. If the force is "roved", as when using an impact hammer for the excitation, the Reference coordinate will be the coordinate location of an accelerometer which remains fixed for all the measurements and the Response coordinates will be the impact locations and directions.

The software will use the measured FRF functions in a process called "curve fitting" to extract the modal parameters of natural frequency, damping, and the mode shape for each mode found.

Modal Analysis Curve Fitting

Curve fitting in the Modal task is a two-step process. First, a table of modal parameters is calculated from one function (depending on the method used). The second step is to process all the measured FRFs to calculate the mode shape.

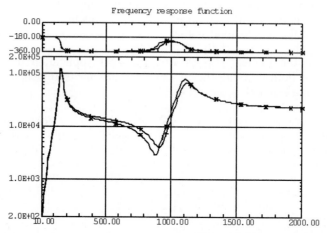

After selecting a parameter extraction method, select a function to identify the natural frequencies. This will usually be the driving point function, where the Reference and Response coordinates are the same. The icon Pick Peaks or Search Peaks will find the frequencies of the peaks. The icon Calculate Residues will calculate the damping and magnitude values for each of the identified parameters. To check the accuracy of the curve fit, the analytic fit based on the extracted parameters is plotted with the original function. The Parameter Table is stored on the Parameter ADF file.

The mode shapes for each natural frequency for all the FRFs are calculated using one of several techniques. Again, you must pick a method such as Move Response (take the peak value) or *SDOF_Polynomial* (use a polynomial fit around each peak). The program will prompt you for the list of coordinates to use from the Function ADF. Values for the mode shape coefficients will be calculated for each of these FRFs. For a "quick look" at the mode shapes, you may want to just use a subset of the measurements taken. For a more accurate shape, you might use a more advanced curve-fitting algorithm and all the measurement points. These more advanced curve fitting algorithms will not be discussed in this introduction, but you can get information on them in the online Help Library.

Model Preparation Task

To display mode shapes, finite element geometry must be created in the Model Preparation task (similar to the Mesh Creation task in the Simulation application). The modes will be displayed as deformed geometry plots or animation in the Post Processing task (again the same as the Post Processing task in Finite Element Analysis).

Geometry is created in the form of nodes and elements, just like in Simulation. In fact, finite element models are interchangeable between Simulation and Test. This will facilitate comparing analytical and test results. The one new geometry concept not as commonly used in Engineering Analysis is that of a "trace line" to display geometry. A trace line is like an element in that it connects nodes for display, but it has no physical or material properties. Either elements or trace lines may be used to display the geometry, but thin shell elements are often preferred, since more advanced hidden line and shaded image displays can be used.

Since the concepts of creating nodes and elements are the same as in Simulation, they are not covered again in this chapter. This material is covered in Chapter 11. The only difference is that mesh generation is not available in Test, since this does not usually make sense for test models.

Post Processing Task

The steps in Post Processing to display the mode shapes are the same as Simulation, Post Processing, as described in Chapter 9. The only differences are that the mode shape is selected from the Shape ADF rather than from an Analysis Dataset, and the display of elements may also contain trace lines.

Where To Go For More Information

This section has outlined some of the capability of the Test application, focusing on function manipulation and modal analysis. For other capabilities, such as data acquisition and fatigue analysis, read the on-line manuals.

For more information on the Test application, see the following manuals in the online Help Library.

> *Help, Help Library*
> > *Test*
> > > *Basic Capabilities User's Guide*
> > > *Signal Processing User's Guide*
> > > *Modal Analysis User's Guide*
> > > *Open Architecture User's Guide*

See also the online tutorials.

> *Help, Help Library, Tutorials*
> > *Test*
> > > *Fundamentals*
> > > > *Quick Tips to Using Test*
> > > > *Introduction to Test*
> > > *Advanced Projects*
> > > > *Exploring I-DEAS Test (Print to use)*

Summary

This chapter has given an overview of the products not discussed in as much detail in earlier chapters of this Student Guide. For a complete listing of products offered by EDS PLM Solutions, select the web page:

> http://www.eds.com/products/plm/

There are also other software products by software vendor partners. There are over one hundred Solution Partners listed on the EDS PLM Solutions web page:

> http://www.eds.com/products/plm/partners/

Chapter 11
Best Practices

Design Strategy

In preceding chapters, we introduced many different modeling tools for the design process. Part modeling can be summarized as three activities:

1. Selecting the basic starting part and the overall modeling strategy,
2. Applying specific features to the model, and
3. Modifying the part.

These three activities are very related. For example, the overall modeling strategy you choose will largely determine which features you will need to apply to complete the model. How you build the model will have a great impact on how easy it will be to modify it. Considering the types of modifications that may be made will lead to a better strategy of modeling a particular part for its intended downstream uses.

Discussing part modeling strategy will serve as a review to tie together many of the concepts presented in previous chapters.

1. Selecting the Basic Starting Part and the Overall Strategy

There are usually many different ways to model any given part. What is the best way to make it? Does it make any difference how you model it? It often does make a big difference. Two parts that may look identical graphically may look drastically different if you display a history tree of each. If one part took 20 steps to create and the other took 200, there will be a big difference in the time it takes the software to make modifications to each. It will also take much less disk space to store the first part.

Some general rules in deciding on a modeling strategy are:

- Model the part with the minimum number of steps possible.
- Model the part in a way that is understandable to others who may use it later.
- Consider the downstream uses of the part.

Some factors that will affect the modeling strategy may be found when you consider your design process. The first question to ask is, are you designing a part, or just documenting a finished design? Often, users look at the solid modeling process as though it were the same as traditional drafting. If the tools are being used as part of the conceptual design, there will be creative freedom and iteration involved in the process, not just transcribing geometry from a 2D drafting or sketch to a 3D computer model.

Sometimes parts may have been designed in 2D drafting a certain way only for convenience based on the tools available. To just copy this design approach into 3D solid modeling may be too constraining for the new types of tools available. For example, a 2D limitation may be to design a part in a way so that it can be dimensioned on paper. In 3D solid modeling, a sculpted surface may make a better design. The 2D dimensioning may not be important if NC machining will be used directly off the electronic files.

Users that try to transcribe drawings into solids can also get bogged down in details such as over- or under-dimensioned drawings, or details that can easily be described on a drawing but do not make 3D geometric sense. If the part is being designed in the I-DEAS software, it is often more convenient to shape the part, but not fully constrain the wireframe geometry for each feature. Adding this detail can be considered part of the third step, modifying the part.

It is important to take a fresh look and ask what are the real constraints in the design. Which mating faces are most critical? Which parameters are the most important? The modeling process should give these factors precedence. These features and parameters should be modeled first, so that later features can use them in a way so that if the key parameters are changed, later steps will be automatically changed.

Someone might look at a part from the point of view of a machinist and start with a basic block of material that needs to be machined away to produce the final part. Someone else might look at the same part and visualize a small starting piece with a collection of features glued on to it. In some cases, one method might show distinct advantages over the other. In other cases, a totally different approach might be far more productive, such as modeling one section of the final part, and reflecting it or using patterns to produce the final part.

Selecting the base starting part will affect the overall strategy of building the model, and may be the most important decision to make. This part most often will be a simple extruded or revolved part. One rule would be to pick the shape that most closely matches the final shape of the part. On the other hand, sometimes it may be better to start with the most important mating surface, or datum of the part. There is no one right answer, but you should consider different strategies.

In some cases, the starting part may not end up in the final part at all. This strategy might be called using a "precursor" part. This part might be used much like an assembly jig that holds things together, but is removed later. For example, to create a sheet metal part, you can create a solid part that has an outer surface to use to create the sheet metal panels.

One of the best techniques is to start your part by creating reference coordinate systems and reference planes at important datum locations. It is preferable to create reference geometry before features rather than to create reference geometry on features that may later be deleted. The easiest way to do this is to use the *New Part* command. Reference geometry created using offset dimensions can be easily modified. Features sketched on the reference geometry will be repositioned when the part is updated. This method has been called the BORN (Base Orphan Reference Node) technique because the base node in the history tree containing the reference coordinate system has no displayable geometry.

In the Assembly task, an equivalent command is *Add Empty Part*. This command combines the steps of creating a reference coordinate system, naming it, and adding it as an instance of the part to your assembly.

2. Applying Features

After you have built the first starting part (which is really not different than any other feature in the part, except that it was done first), the rest of the process is simply adding features to it.

Some general rules to consider are-

- Use the minimum number of steps.
- If there is more than one way to apply a feature, do it the simplest way.
- Consider the order of features.
- Consider the associativity of how features are created.
- Use symmetry to create half the part and then reflect.

In all of this discussion, it is assumed that the reader has a basic understanding of the concept of part history. Without this understanding of the history tree, it is impossible to understand what we mean by the general rule "use the minimum number of steps."

A common occurrence with new users is to create parts containing extra steps in the history tree that do not show graphically. This might happen, for example, if they cut or protrude in the wrong direction. Instead of modifying or deleting this step, they did it over again until they got it right. The first step, however, unknown to them, is still recorded in the history tree, and is wasting space in the database. These extra unused features will make it very difficult for someone else to understand.

With a proper understanding of the history tree concept, you can understand that you should not use a complex part only for the purpose of cutting a simple cutout in another part, because the resulting part will contain the complete history of both branches. It would be better to re-model the cutting interface using a simple part or surface, just to keep the history tree simple.

When complex features or sections of parts are modeled separately and then joined, the history tree has a "bushy" look instead of just a long "vine." This can have some advantages in editing and replaying a part's history. Some sophisticated users manage the history tree as though it were an assembly hierarchy. Individual users are responsible for modeling different branches of the tree.

If a part has repeated features, you should look to see if the features can be created in one step, if a pattern can be used, or if symmetry can be used. Finite Element Analysts tend to look for symmetry, and model only the half (or quarter, etc.) required. Designers tend to overlook symmetry because they tend to look at the modeling process like the manufacturing process.

Another technique is to look for ways to combine features in one step. For example, if four holes will be drilled in the same face of a block, these should be cut as one step, not as four separate steps. This will not only minimize the number of steps in the history tree, but it will also make the part easier to modify, because it will be easier to apply constraints to the wireframe geometry of the feature.

Sometimes it may look as though two features may require two steps, since they are cut into two different faces of a part. Users tend to only sketch wireframe geometry within the boundaries of the face being sketched on. This face should be viewed as an infinite plane that can be used for sketching. The boundary shown gives you some geometry to align to, but is not a boundary you must stay within, nor does it limit you to only aligning to this geometry. An often overlooked command option is *Focus*, which will allow you to project points from anywhere on the part onto the sketch plane. Points projected using this option remain associative to the original points, so this method also helps to make later design changes easy to make.

Look at features carefully to find the easiest way to create them. For example, always try to use an extruded or revolved shape if possible. Sometimes a feature that looks like a loft or a sweep may really be an extrusion with a draft angle.

Use a sweep along a 2D path before choosing to use a loft with multiple sections. If using a loft, use the fewest number of sections possible. Using more may mean more trial and error to get smooth surfaces without bumps.

Every feature should be attached with some kind of relation to the part. If you just join a feature with the relations switch turned off, and then make a change to the part, the feature will just remain where it was in space, with no associativity to the previous steps. Features that are joined using the *Cut*, *Join*, or *Intersect* commands should be positioned using these relations if you want to be able to make associative changes to earlier steps. Features that were sketched in place should be tied with variational constraints and dimensions to the edges of the sketch face. If the geometry you want to relate the feature to is not on the sketch plane, you can use the Focus option to project an associative point, and then dimension to it.

Give important dimensions a meaningful name so that they will be easier to modify later, or easier to recognize if you want to relate other parameters to them.

Although a general rule is to include the most geometry possible in each feature, if large topology changes will be made, some features are better left as a separate step. For example, if the dimension driving the location of a hole may vary enough to sometimes put the hole inside a face, sometimes outside, or sometimes half in and half out, this hole should be cut as a separate feature, not as part of the same wireframe geometry defining the face to be extruded. If not, the section definition may become ambiguous. If the hole moves outside the section, it is not clear if this is to be a hole or an island of material. It would be better in this case to define a separate hole feature. There is no problem with topology if the hole is inside, outside, or half inside. In all cases, it remains a cut operation.

When creating the section for a feature, *Section Options* control the chaining. Remember that these options will also be reapplied with the part is modified. Setting the options so that the section can be chained in the minimum number of picks usually also results in the most reliable replay of the part when the geometry changes.

3. Modifying the Design

It is not a trivial task to take a complicated part model created by someone else and make design changes to it or make other changes for a particular application. This task can be made simpler if the creator of the part has modeled it in a logical order, and anticipated the types of changes that might be made to it, and the types of downstream uses of the part model.

One consideration of the downstream uses of a part is the range of variation expected in each parameter. If you expect large design changes to be made, it will be worth the time to set up variational constraints to fully define the design intent of each section. While it is a major time savings in the design stage that parts do not have to be fully dimensioned, at some point as the design evolves, you should go back and add the dimensions and constraints to each section so that they are fully constrained. This will guarantee the most predictable results when large dimensional changes are made. It will be easier for the original creator of a part who understands each feature of a part to add these constraints than for someone else to understand the design intent later.

Although it will make a part history easier to understand if the maximum amount of detail is added in each section used to define each feature rather than add extra features, having some of the features as separate steps will make them easier to suppress for some downstream applications. For example, often small fillets will be ignored by the finite element analyst. Also, internal radii might be the result of the cutter used in the machining process. The machinist would rather only work with the required surfaces, not those that are intended to be artifacts of the machining process. A common mistake is to add more detail than is required. Features that may be required for some applications but not for others should generally be applied last, to make them easier to suppress.

There are several ways to make simple dimensional modifications. If you pick the part, the modification options are Show Dimensions and Dimension Values. The first option will display all the key dimensions on the part. Note that this will not show dimensions that are defined by an equation to another dimension or equation. It will only show the independent driving dimensions. If the part has been properly modeled, any user will be able to modify these driving values with predictable results. If dimensions are named, it will make it easier to locate them and modify them using the Dimension Values option, which places all of the dimensions in a table.

If you select a feature, there are more options available to make modifications. Show Dimensions and Dimension Values options are available, but they only show dimensions relating to the particular feature. Other options are Feature Parameters, which allows you to modify any of the original feature parameters defined on the Extrude (or other) form used to define the feature. This option should be used, for example, if you extruded a feature in the wrong direction.

You may use the *Modify, Wireframe* command to "open" the wireframe geometry as it was when the feature was sketched. This option allows you to change dimension values, constraints, curves, or redefine the section which traces around the curves to define the feature shape.

You should be aware of how associativity works if you make changes to the defining section. Each curve used in the section has a label number. These labels can be turned on using *Display Filters* if you want to see them (which you normally won't). When the section is extruded into a protrusion or cutout, each of these curves generates a surface which also has a label number. If you later attach a feature to one of these surfaces, it is associated to the surface label, which is associated back to the original curve label. If you make a drastic change to the section, such as deleting it and re-defining it, you will break this associativity of the dependent feature. The feature will not go away, but will not be associative to part changes until you do something to add relations back to define how it is to be positioned.

For this reason, it is usually better to make the minimum change to a section. Instead of deleting it completely, use the *Modify* command, pick the section, and remove or add curves to it. Be particularly aware of curves on the section that may have features tied to them.

Other options available with the *Modify* command when a feature is selected are *Suppress Feature*, *Replace Feature*, *Add Relations*, and *Delete Relations*. *Add Relations* can be used to build associativity back if you had deleted it or broken it by making a section change as described above.

Some other options are available when you select a feature that may not be quite as obvious, since they are not options under the *Modify* command but involve using other commands. For example, with a feature selected you may use the *Delete* command to completely delete a feature (as opposed to temporarily suppressing it). You may also move a feature if it was joined without using relations.

Another option in menu commands is to *Insert* a feature or *Reorder* a feature. Sometimes it may be necessary to change the order in which features are constructed. This also allows taking complicated features from existing parts and using them on other parts.

It was already discussed above how to avoid losing associativity when modifying sections used to define a feature. This same loss of associativity can also happen when a change in dimension causes edges to appear or disappear where features are attached. For example, fillets are a feature attached to edges. If a topology change occurs, there may be more edges or fewer edges that you want included in the fillet. If this happens, do not just make a fillet feature again, or you will end up with a part history that nobody will be able to understand. The correct approach is to select the fillet feature and modify the edges picked. When the graphical history tree is displayed, any features such as this that have an error on replay will be displayed in a different color to help you find the error. Although these part modification problems can be fixed, in some cases it may be possible to anticipate potential problems by considering the range of possible future dimensional changes that might be made.

Summary

This chapter gave an overview of some things to consider in deciding on the modeling strategy before starting to build a part model. These general rules should be kept in mind before starting to build a part model.

⟨💡⟩ PARTING_SUGGESTIONS

- Start parts with a coordinate system using the *New Part* command You may then sketch on faces of this coordinate system. You may also use the assembly command *Add Empty Part* to create a new part.
- The most important feature is the first (base part) created. The selection of this feature often determines the entire part modeling strategy.
- In general, the method that uses the fewest history steps is the best.
- Name features and dimensions to help others modify your parts.
- Use reference geometry to locate features rather than to build features on other features that may be deleted later.
- If the same geometry is used in multiple features, create reference curves from this geometry as part of the first feature.
- Avoid cutting a part with a part that has a large history. Instead, extract the feature needed from the cutting part and use this extracted part as the cutter.
- If it is possible to create a feature using simple techniques such as extrude with draft and applying fillets, it is usually preferable to do it that way.
- Use "bottoms-up" surface creation and stitching into solid parts only as a last resort, such as when forced to use geometry from external CAD systems. It is preferable to model parts using only construction operations.

Where To Go For More Information:

For more information on best practices, see the articles in:

Help, Help Library
    Design User's Guide
        Part Design
            Design Strategies and Concepts
        Assemblies - Creating and Managing
            Constraining Assemblies
                Constraining and Dimensioning Instances
                    Best Practices for Constraining Instances
        Assemblies - Working as a Team
            Improving Model Performance
                Best Practices for Large Assemblies
    Data Management User's Guides
        Project Management User's Guide
            Project Planning
                Best Practices - Sharing Data

Search for the following terms:

Associativity
Practic*
Strateg*

Tutorial: Best Practices

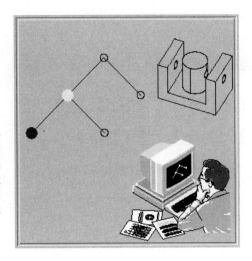

There are usually many different ways to model a part. Although different methods may result in parts that look the same, the differences will be more apparent later when changes are required.

Before you start:

1. You should have a good working knowledge of creating and manipulating parts.

2. You should have completed Chapters 2-4 in this Student Guide.

What you should learn:

1. How to use reference geometry.
2. How to use patterns.
3. How to use good practices with part history.

Tutorial Instructions.

Follow the instructions in the tutorials below.

Help, Help Library, Tutorials
Design - Part Modeling
Fundamental Skills
Using Reference Geometry
Creating Patterns
Modeler Troubleshooting

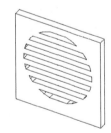

These tutorials will give you a deeper understanding of the history tree. They also show techniques to minimize the number of steps in the history tree and include associativity between features.

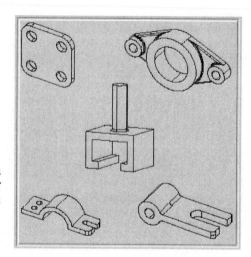

Tutorial: Creating Basic Machine Parts

This tutorial presents a collection of simple parts for you to practice modeling on your own.

Before you start:

1. You should have a good working knowledge of creating and manipulating parts.

2. You should have completed Chapters 2-4 in this Student Guide.

What you should learn:

1. Practice looking at each part and deciding a modeling strategy.

Tutorial Instructions.

Follow the instructions in the tutorials below.

Help, Help Library, Tutorials
Design - Part Modeling
Advanced Projects
Creating Basic Machine Parts
Modeling a Turned Part

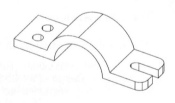

In the part at the right below, can you see a better way to cut the holes as one feature instead of two as shown in the tutorial? (Sketch on the back face, and cut all three holes at once.)

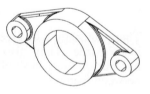

In the other parts, think about good modeling practice. Don't just follow the tutorial steps.

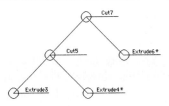

Workshop 11: Best Practices

This workshop will let you experiment with creating some parts on your own, to think about the strategy of how to begin modeling different parts.

Before you start:

1. You should have a good working knowledge of creating and manipulating parts.

2. You should have completed Chapters 2-4 in this Student Guide.

After you're done, you should be able to:

1. Propose alternate ways to model parts.
2. Analyze which approaches lead to the best practices for your particular applications.
3. Decide on a best modeling strategy.

Workshop Instructions:

Model a part like the one shown below using simple extrusions. This part should have a history tree as shown.

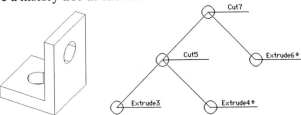

Can you model the part below in the same number of steps?

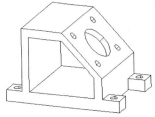

Create this part, then test what happens when you modify some of the dimensions. Does the model contain your desired design intent?

Automotive Door Hinge

You may have modeled this automotive door hinge part in a class or another tutorial. Try the alternate methods to follow, showing different methods for modeling this part.

Method 1 - Start with the side showing the most detail

Model the part as shown below. This strategy is to start with the side that has the most detail. The method results in a history tree with four leaves:

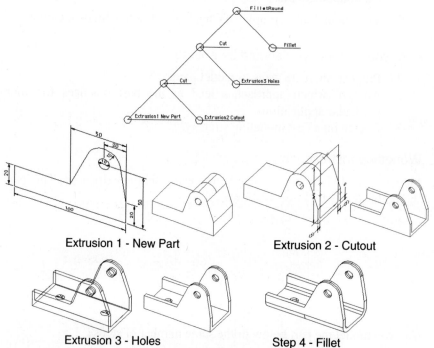

Extrusion 1 - New Part   Extrusion 2 - Cutout

Extrusion 3 - Holes   Step 4 - Fillet

Next, as you model the hinge using three other methods, see how these methods compare.

How many leaves in the history tree does each take?
How easy is it to change:
    Sheet metal thickness, bend radius, distance between sides?
How easy is it to suppress the bend radius?

Method 2- Shell

Since this is a sheet metal part, it has one thickness, and is easily created with the Shell command. This will make it easy to modify the thickness with one parameter without the necessity of matching dimensions. Try building this part with the same first extrusion, then fillet the lower edges, then shell, being careful to delete the surfaces you don't want. As a last step, cut the mounting holes.

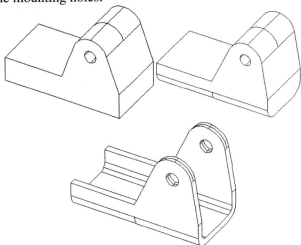

What does the resulting history tree look like? (Hint: It should also have four leaves- extrude, fillet, shell, extrude.)

Method 3 - Extruded channel section, cut away

A manufacturing approach might be to start with an extruded channel section (with the fillet already on it). Cut away pieces and the hole. The last step is to cut the mounting holes.

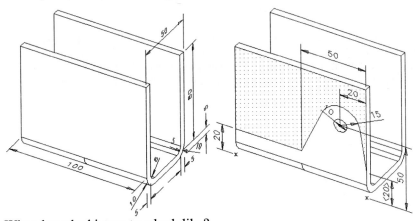

What does the history tree look like?
(Hint: It should only have three leaves- all extrusions.)

Method 4 - Intersect

The intersection operator often leads to an efficient part model, but is often overlooked, since is not as physically intuitive as cutting and protruding. Build the model with the same first extrusion as in method 1, then intersect the channel section. Lastly, cut the mounting holes.

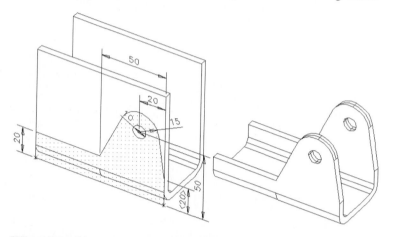

What does this history tree look like?
(Hint: It should look very similar to method 3, but with only three leaves.)

Conclusion

Which method is the best? For the minimum number of history steps, methods 3 or 4 have an advantage. If however, you might want to suppress a given feature, such as the bend radius, it would be harder with these methods, since this would require modifying the wireframe section to remove the fillet. Methods 3 and 4 also require using the *Focus* option and *Stop at Intersections*. Method 2 is easy to perform, except the shell step requires multiple picks to define the faces to delete.

There is often more than one good method, and which is best might depend on what you will do with the part model. Will the manufacturing department want each feature as a separate step? Will the analyst want to suppress fillets or holes? These are all considerations to take into account, as well as modeling simplicity.

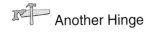

Another Hinge

Try modeling the automotive hinge parts shown. Use your design judgment for dimensions not given. Although similar to the previous hinge, the additional detail will add complexity. Try to plan a design strategy to model the parts with a minimum number of history steps.

Hints:

Use extra construction lines to dimension the angles as shown.

Sketch just some of the lines, then add all the required constraints to fully constrain them.

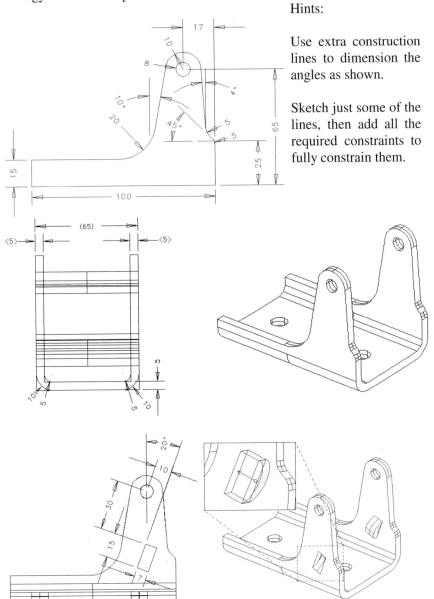

Hint: cut the rectangle all the way through, then protrude arcs from the side of the hole to create the door stop feature.

 Door Side of Hinge

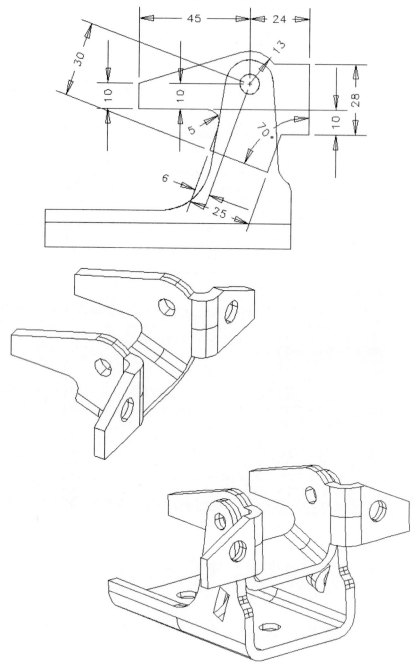

Exit when you are finished.

 ...

Chapter 12
Collaboration

Concurrent Engineering

The philosophy behind I-DEAS is to encourage concurrent engineering design by offering an integrated set of design automation tools. The software can be used by individual users working alone, not sharing data with others, but the real advantages occur when these tools are used together, allowing closer communication between different disciplines working collaboratively on a design project.

Two important aspects to effectively implementing concurrent engineering are first, how to organize your data, and second, how to define your process. The first issue deals with the static picture of how data is organized into projects, libraries, and catalogs. The second issue is the dynamics of how the data changes with time. Who originates design concepts? What design steps are followed between initial concepts to final released part models and drawings? Who can make changes at different stages of the design? How are members of the design team notified of changes?

Organizing Data

The first issue is how to organize your data. Depending on the size of your company, you might organize your data differently, but the basic issue is the same. How do you want to organize your work using the different containers that I-DEAS Data Management provides? These are projects, libraries, catalogs, model files, and bins within the model file. The first step is to understand how these containers work, to understand how you can best use them.

Because I-DEAS data management keeps tracks of your data, the physical file locations is a secondary issue, but it is important that these decisions be made before extensive projects are started, because it will become harder to reorganize your data as the amount grows. The locations of files will be discussed later in this chapter.

Defining a Process

In order to use these tools in a concurrent engineering environment, models need to be shared between users in a controlled manner. What are the design phases? Who is allowed to make changes during each phase? These questions become more important as the size of the design team grows. Among a small work group, informal agreements can be made. In a larger team, it may be necessary to configure each project to define design states, user roles, and access permissions in each state to enforce a pre-defined design process.

Interoperability

Interoperability allows you to collaboratively share models between the Solid Edge and Unigraphics software of EDS PLM Solutions.

Working with I-DEAS™ Data Management (IDM)

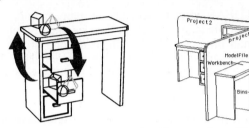

This desk above was used in Chapter 1 to illustrate the concepts of the desktop and bins in the model file. The model file is a container holding the workbench and the bins (and other things).

A further expansion of the picture of I-DEAS data management (IDM) might look like the second picture. Projects are containers for model files, libraries, and catalogs.

This chapter will discuss more of the philosophy of how to effectively use these containers to organize your data and your process.

Organizing Projects

In the illustration on the previous page, each office cubicle (project) is separate from the others, but nothing prevents you from walking around the corner to borrow something from the library of a neighboring cubicle (project). The same is true in I-DEAS Data Management. You can check in/out parts to/from other projects. You will have to do an extra step of setting the context to include these projects in Manage Items to find parts in these other libraries. You need to decide then, do you want to use more projects with less data in them, or fewer projects, with more items in each?

The best way to organize I-DEAS projects depends on many factors, such as the size and organization of the team as well as the size and complexity of the shared data. In a small project with a small team, one project for each product may be the most efficient way to organize your data. This will become unwieldy as the numbers of team members and the number of shared data items grows.

Some companies use an organizational strategy where users have their own projects for initial concepts, and other projects for shared team information. Other projects yet may be used to organize standard parts that are used in many products.

Another strategy is to organize projects in parallel with the company's design process. For example, final released designs may be placed in a separate project to give factory floor personnel access to only this information, but not to in-process information.

The organizational structure of stored data and naming conventions should be defined at the very beginning of a design cycle. This will help provide access to those who need the data, make it easier to prevent access from those who should not have it, and also make it easier for the computer system manager to back up shared data.

Use of Libraries

Libraries are the primary mechanism used to store parts, drawings, and assemblies to allow data sharing among a team. Users should also think of libraries as the main storage location for their important models, including the final copy and optionally the history of earlier versions. This is true whether they plan for other team members to share this data or not.

When users share data, the library offers a locking mechanism to prevent them from making conflicting modifications to a part at the same time. Many users can "reference" a part out of the library, but only one user at a time can "check-out" a part to be able to change it. The data management system in the software keeps track of who has a part checked out, and lets you update your models that contain parts taken out of the library as reference copies.

Libraries are part of a project. To share a common library with another user, either select the same project name when you both enter the I-DEAS software, or select this project when getting the part from the library. Libraries are designed to keep track of parts and assemblies as the product evolves.

When you get a part, assembly, or drawing out of a library, you have three choices:

Check out	Check out a part, assembly, or drawing to make changes. Only one user can check out a part to make changes at a time. After you make changes, check the part back into the library.
Reference	Get from the library to use for reference only. You cannot make changes, but you can update your drawings, assemblies, or simulations, etc., that use this part if the originator changes it. You have the option to get a specific version rather than the latest version. You cannot modify parts accessed for reference only.
Copy	Get an unassociated local copy. This option is useful if you want to make "what if" changes to a part, but do not want to make these changes permanent. This option should be used if you want to perform finite element shape optimization, because optimization makes changes to the part geometry. With a copy, you have the option to be notified of changes.

A library has utility even for single users. One use is as a backup copy of your parts. If something happens to your model file, such as running out of file space, the parts previously placed in the library will still be safe. The library is inherently a safer place to keep data, since each part is stored in a separate file. In a model file, one bit out of place could make the entire file unreadable. In a library, one bit out of place may damage one part, but not all of them.

Another reason single users as well as team members should consider using libraries is that when the part is written to a library, some clean-up and checking of the data is done at that time. This means that parts that have been successfully checked into a library are less apt to contain hidden errors that may cause problems later. If you have problems checking a part into a library, it could be due to the fact that there are errors in the model. In this case, don't assume that the problem is with the library, it could be that something is wrong with your part model. You should try to find out what is wrong and fix it.

Library Versions and Revisions

The library also has a mechanism to keep track of versions of a part. If you check-out a part, modify it, and check it back in, the library will contain both versions. This is not true if you only store your parts in a model file. There, you would have to manually store different versions with a different name, since the model file does not have the automatic mechanism of storing old versions. Other users can get a copy of a part out of the library. If they check this part back in to the library, it will not be stored as a new version of the original part, but will be a new and different part.

Version numbers are automatically incremented each time a part is checked out, changed, and checked back into the library. The user can assign revision numbers to specific versions of a part. Revisions are used to keep track of released versions of a product, where the automatic version numbers keep track of day-to-day changes in the design. While version numbers are automatically assigned by the software, users can assign their own revision numbers.

Catalogs

Catalogs are very similar to libraries, but are intended to store parts that do not change. Catalogs only maintain the latest version of each part. Parts can, however, be placed into a catalog as a parameterized part that lets the user change key parameters. Parts can also be placed into the catalog as a part family, which contains specific tables of values of the variable parameters. A part family would be used to allow users to get specific configurations of standard nuts and bolts, for example.

There are three kinds of catalogs– part catalogs, feature catalogs, and section catalogs. Part catalogs contain complete parts either as different standard parts, or as a "parameterized part," which can be changed into any number of different combinations of its defining parameters. Feature catalogs store parts that are used as cutting or joining features. Section catalogs contain section shapes that can be used to create or modify parts.

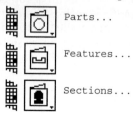

Parts...

Features...

Sections...

Items and Metadata

In data management systems, information is stored about data files. For example, the data management system keeps track of the names of the files and where they are located, the type of information in files, and other useful information such as who created the files, and when they were created. The data management system can be described as storing "data about data." This concept of "data about data" is called "metadata."

In IDM, this metadata is stored for each "item" that IDM tracks. An item, then, is the fundamental unit of data management by IDM. In the metadata files where this information is tracked, information is stored about the item name, item type, part number, version number, relationships, and other attributes.

For an example, look back at the illustration of the part definition in Chapter 3 of this Student Guide. Notice that the "Metadata" attributes include things like the name, part number, version number, revision number, create date, creator, etc. This data is fundamentally different than the other attributes of the part such as its appearance color, history, and topology. All of these other part attributes are stored with the part, and make up its definition. The item attributes are things that other team members need to know about to even know about the existence of your part. These item attributes are metadata that should be shared with the rest of the team. For this reason, this data is stored in a central place where all team members can access it.

There are many different types of items that are tracked by IDM. The concept of a part was given only as one example. Other item types include model files, drawings, FE models, NC jobs, external data files, etc.

In the case of a part, you assign the part name, which is the same as the item name tracked by IDM. When you create a drawing in Drafting, you are prompted for an item name. This allows the IDM item to have a different name from the physical filename.

One of the important functions of IDM is to keep track of relationships between items. For example, an FE model is dependent on a part model. The software must not let you delete the part if the FE model needs it, unless it deletes both items. Although the software automatically creates the required relationships, you can add your own relationships between items. For information on this topic, search for the word "relationships" in the Help Library in the Data Management User's Guide.

Managing Metadata Items

There are several commands that let you view metadata items. The *Manage Bins* command displays items in a model file. You can select the name of an item, then use the *Details* menu to display more information such as the creation date of the item. You may also delete items here.

The Manage Library icon displays metadata about items in libraries.

 In versions before I-DEAS 8, the *Manage Items* command displayed every item in libraries and model files. In I-DEAS 8 or higher, use *File, Manage* for information about model files. Use *Manage Libraries* for information about things in libraries.

On page 44, the correct way to delete model files you no longer need was shown. It is important to use software commands to delete model files rather than deleting them at the operating system level. When this is done, the "item" describing this file will be removed from the project metadata file. This is true for any of the items tracked by IDM. This will keep IDM synchronized with the actual files on the operating system.

Configuring Projects

There is often confusion over another capability in I-DEAS called I-DEAS Team Data Manager (TDM). TDM provides two main functions beyond the basic capability of IDM. First, the project manager can configure the roles of each individual working on a project and restrict data access to certain individuals at certain states of the design. The second main function of TDM is to support active E-mail notification of design changes to required individuals.

Many users get confused over where IDM ends and TDM starts, and refer to both as TDM. TDM includes the functions inside the "Configure Project" command. TDM does not provide any new capability to what can be modeled or managed within I-DEAS; it only helps control the flow of information among a team.

Configuring projects is beyond the scope of this discussion. For more information on this topic, look in the online Data Management Project Manager's Guide.

Where is Your Data Stored?

Now that you understand the difference between your data and metadata, you can understand that these are stored in different locations. Your data is first created in your model file. When you name a part, the software creates the metadata describing that part. For example, the part is fully identified by a name, part number, and a version number. While the part itself is in your model file, the metadata about that part is stored in a central project file, which will be described more fully below.

When you check your part into a library, the metadata is still stored in the project file, but the part data is now stored in the library file. A copy may also be kept in the model file, if you asked to keep a copy. We will next look at more detail at what is in these files, and where they are actually located in the file system.

Master Metadata File

At the highest level of data management in I-DEAS, there is a master file named:

    z_master.imd

that keeps track of all the projects. In a typical installation, this file is located in the directory:

    .../team/master

Where it is located on your computer system was a choice made at installation. (This will be discussed more later.)

Notice that the extension on this file is ".imd." This stands for "I-DEAS Meta Data." This file contains a list of all the projects that have been created. When you start I-DEAS software, the list of existing projects that you can choose from comes from this file. If you enter a new project name, that name will be stored in this file.

You can get information about the projects in your data installation using the Manage Projects icon. This information is stored in the Master Metadata File.

Project File

Each project recorded in the master file has its own metadata file to keep track of project information. This allows multiple users logged in at the same time to share this same file and access each other's metadata stored in this common file. In a typical installation, the name of this file is:

```
.../team/projects/z_xxx.pmd
```

The "xxx" in the name shown will be replaced by a unique string for each project file. Notice that the extension on this file is ".pmd". This stands for "Project Meta Data."

This project file contains the metadata for every item created by any user who selects this project when entering the I-DEAS software. It also contains information about any libraries or catalogs associated with the project.

You can look at the contents of your projects using the Manage Items icon.

Libraries and Catalogs

Libraries are used for data that changes as a design evolves, and catalogs are used for standard parts. From a data management perspective, they are the same.

The name of libraries and catalogs associated with a project are stored in the project file shown above. Note that a project can contain multiple libraries and catalogs, but each library and catalog exists in only one project that "owns" it. You can open a library or catalog belonging to Project A from Project B, but the metadata about the library or catalog is still owned by and stored in the project file for Project A.

Library or catalog metadata is not stored in a separate file, but is contained in the project file. Although the metadata about libraries and catalogs (containing the name of the library and the names of library parts) is contained in the project file, the actual library parts are not stored there. They are stored by default in the directory:

```
.../team/shared
```

Each part stored in the library will be stored as a separate file in this directory. The filename will be a unique string assigned by IDM with an extension ".prt".

Model Files

Users should think of model files as their own local workbench, not as the central storage location of data. This concept may be described as "private data." Other members of your project may examine the metadata about items in your model file, but they can't directly get access to the items themselves.

Private and Shared Data

Model files are generally described as containing "private data", and libraries as containing "shared data." As mentioned above, other users in the same project can view the metadata about what is in your model file, but they can't directly get the data.

As you gain a basic understanding of the "underlying machinery" of how the data is stored and managed in IDM, you can see that libraries have more utility than just using them for shared data. You can also understand how problems might be caused in IDM by trying to share data in other ways. The types of problems that can occur will be discussed more fully a little later.

Temporary Scratch Files

Both model files and metadata files use "shadow" scratch files while you are running the software to store transactions that have taken place since the last Save command. These files have the extensions of ".smd" for "Shadow Meta Data" and ".sf1" and ".sf2" for model file scratch files.

There is an important concept to notice here! In both the cases of model files and the metadata files (master file and project file) the latest transactions are not permanently written to the file until you give the Save command. For example, if you check a part into a library, other users will not see it until after you give the Save command. This is why when you perform a library check-in operation, a form will ask you if you want to save so other users will be notified of the operation.

A preference switch allows the software to automatically save after library operations. Leaving this switch on will keep your model file synchronized with the project metadata.

A general rule is that you should save often.

Avoiding Problems

Problems can occur when you try to directly copy, delete, or rename files using operating system commands. For example, if you copy a model file and overwrite another file with the same name, IDM will think the model file is the same one because the name is the same, and the file location is the same. IDM will get confused, however, when the items in the file don't match those in its metadata file. This again illustrates the importance of understanding the basic machinery behind how the software works.

What is the proper way to give another user a copy of the parts in your model file? The preferred way is to place the parts in a common library. A second option is to export an archive file, which is an ASCII file containing the contents of the model file.

Another reported confusion happens when the user issues the "Save As" command to make a new copy. What if the original model file had a part checked out from the library? IDM could not allow two model files to both have the part checked out, so it must convert the library status of one of the files to "copy" instead of "checked-out." Again, what would prevent this problem? (Use libraries as the central repository for information.)

A problem that can happen is that a user checks out a part, saves his model file, and then goes on vacation for two weeks, or his computer crashes while he has a part checked out. The model file has a lock on that part, and others will not be able to modify it. What do you do? The best answer is to prevent this from happening in the first place, and don't keep parts or assemblies checked out in your model file. A second answer is to get the password of the user and log in to the same account. A last resort is to override the lock the user had on the file. This is not done without consequences, however, since when the user comes back, his model file will have the part or assembly as a copy, and changes made can't be checked in as new versions. The item can be unprotected using one of the manage commands and using: ◆ *Modify, Action, Unprotect*.

Now that you understand how the metadata files track information in IDM, you also understand how certain operations can confuse the system. For example, it was mentioned above how you should delete a model file by letting the software delete it, rather than delete it from the operating system. Otherwise, IDM is still trying to track the model file and all the parts in it. Similarly, making a copy of the model file, and opening it in the same project can cause problems, because all of the items have the same name.

For more information, see the section "Troubleshooting I-DEAS Data Management" in the Data Management User's Guide online. You may want to print this article to keep as a reference.

Data Installation

I-DEAS has a concept of a data installation, which is separate from the software installation. To the user, a data installation is all of the projects that are visible to them in the software.

Two major decisions must be addressed as part of the installation of I-DEAS: *1.) Who will own the files that I-DEAS creates? and 2.) Where will I-DEAS data files be stored?*

The install program will ask question number 1. The default answer is that the account "ideasadm" will own all I-DEAS data files. This is recommended in most cases, because it will prevent users from deleting files tracked by I-DEAS Data Management. Rather than deleting files using operating system commands, the users will use commands and utilities in I-DEAS to delete items. This will prevent users from deleting metadata files, and from deleting files that I-DEAS Data Management "thinks" are there.

Another option to file ownership is to use "user ownership." This option is recommended only for training installations, universities, or standalone installations where data is not normally shared between users. This will allow users to easily delete all of their own I-DEAS data files at the operating system level after a class or project is finished.

This decision is made during the installation, and applies to all users. You cannot have some users own their own data files, and ideasadm own others. Therefore, the implications of this question should be carefully thought out before the installation. There is, however, a utility to allow the system administrator to switch the installation from one type to the other without doing a complete re-install.

This topic is known as "Secured File Access Model" (SFAM). For details about this topic, look in the Installation Guide. For a more complete discussion of installation site planning, see the manual: *Site Planning and Implementation Guide*.

Location of Metadata Files

We mentioned earlier that there was one master metadata file. The "world" of projects known to the one master file is called a "data installation." By default, there is one data installation for each software installation. We will discuss below how it is possible to divide up the installation into multiple data installations. It is also possible to have multiple software installations share a common data installation.

An installation decision is where to locate I-DEAS metadata files. If ideasadm owns the files, they must be located in a directory where the ideasadm account has read/write privilege. They should also be located in a directory owned by ideasadm. If the user owns the files, these files must be located in an area where the user has read/write privilege, or the user will not be able to run I-DEAS at all.

The directory locations where I-DEAS files will be written are defined in the file:

```
/ideas_msx/ideas/.ideas_param
```

Depending on the version of the software, this file may be named .ideas_paramX, where X is the version number. On Windows, the file does not include the "." in the filename.

If you are not familiar with this file, you should take the time to look at it in detail. This file is like the central wiring switchboard for I-DEAS. It contains the system default user preference settings and defines where files will be stored. This file is full of comment lines that self-document the functions it performs.

For data management, four lines are particularly important.

`Team.MasterId`: An integer number, unique to a data installation.
`Team.MasterDirectory`: Location of the master metadata file.
`Team.ProjectsDirectory`: Directory containing the project files.
`Team.SharedDirectory`: Directory where catalog and library part files will be stored.

The variable "Team.MasterDirectory" points to a directory that will contain the file z_master.imd. This file is the master file containing the names of all the Projects you see when you start I-DEAS. Each project has a file stored in the Team.ProjectsDirectory that contains the items tracked in that Project. For example, a Project may contain catalogs and libraries. The parts in these catalogs and libraries are stored in the directory Team.SharedDirectory. Default locations for all of these files are defined in the .ideas_param file, but you may want to change these locations as part of the installation. For example, a common practice may be to locate these important files on a central disk used to store sensitive company data.

There may be times when you want to break your users up into separate data installations that can't "see" each others projects or anything within them. This concept is called "multiple data installations." This may be desired, for example, if a company has a small group working on a secret project that the rest of the company should not know about. This concept also applies in a student environment, where students are working on individual assignments, rather than on cooperative projects.

To create multiple data installations, individual users can access different versions of the .ideas_param file. The standard way to do this is to define the variable $IDEAS_PARAM in each user's .profile file. When I-DEAS starts, this variable will point to a directory to find a copy of the .ideas_param file other than the system-wide default copy.

.ideas_param File

There are other ways to redefine the parameters defined in the central .ideas_param file. When I-DEAS starts, it reads this information in two different locations in the following order:

---

1. The central .ideas_param file
   (or a copy pointed to by $IDEAS_PARAM) is read.
2. The user preferences are read.
   (These are stored in the Team.MasterDirectory).

Importing a Model File

If you want to copy a model file into another project, what the software normally will do is alert you to the fact that it is "importing" the model file, rather than creating a new one or opening an existing one.

What has to happen when you "import" a model file into a project? The answer is that the software has to examine the contents of the model file and add metadata items for each of the parts in the file. That is what it really means to the software to "import" a file.

For more information on this topic, search on the word "import" in the Data Management User's Guide in the Help Library.

Adopting Files

Another related operation is to adopt files. This allows you to let IDM track non-I-DEAS files. When you adopt a file, the software creates a metadata item to track the file and makes a copy of the file. You may then want to define relationships between other items in IDM.

For more information on adopting files, search on the word *adopt* in the *General I-DEAS Administration* user's guide in the online Help Library.

▶▶▶ WWW, Images, and VRML

Another way to share data as a team is using the World Wide Web. I-DEAS has several new capabilities for creating 2D and 3D images for the Web, linking to the Web, and sharing information over the Web.

One link to the Web from I-DEAS is the *Help*, *Web* menu, which provides links to the SDRC Web or to your own internal home page.

The print utility has the capability to create JPEG and other files from picture files. Other screen capture utilities can also be used to capture bitmap images from the I-DEAS Graphics window to create images to be displayed on the Web.

Another capability in I-DEAS is to add notes containing URL addresses to parts. The icon *World Wide Web* (found in the *Info* stack of icons) can be used to start the browser and link to the URL address. For more information, search for "URL" in the Help system.

The report writer option found in many I-DEAS commands now has the capability of writing HTML output so that reports can be viewed on the Web. To find documentation on this capability, search for "HTML."

The *File*, *Export* format can export VRML files of parts and assemblies that can be displayed as 3D images with browsers capable of displaying VRML files. For details about this capability, search in the *Geometry Data Translators* user's guide for "VRML."

Summary

This has been a brief summary of ways to work as a team in I-DEAS. This chapter has also outlined how I-DEAS Data Management works. Users are warned that the previous discussion may be somewhat out-of-date, or may be an incomplete summary. For a complete description for your version of the software, see the user's guides in the online Help Library.

Where To Go For More Information

For more information on data management, see the articles in:

Help, Help Library, Bookshelf   or   *Help, Manuals, Table of Contents*
    Data Management Guides
        Data Management User's Guide
            Data Management Basics   or   *Team Design Basics*
            Working with Model Files and Their Items
            Working with Libraries (and Library Items)
            General Item Management
            Working with Catalogs
            Working with Projects
            Troubleshooting I-DEAS Data Management
        Project Management User's Guide
            Project Planning
            Setting Up Projects
            Sharing Data During Design
            Sharing Library Items With Other I-DEAS Installations
            Getting Information About Projects
            Managing Configured Projects

If you are a system manager or project manager, you should also see:

Help, Help Library, Bookshelf
    General I-DEAS Administration

⋅̣̇⋅ PARTING_SUGGESTIONS

-Don't try to manipulate model files or other files from the operating system.

-Store parts and assemblies in libraries both for team use and for your own personal use.

-Let the software keep track of where the files are stored. You can find files by various attributes using software commands.

-Save often.

-Save whenever the software prompts you to.

-Avoid *Save As*. If you have parts checked out in the model file, they will be converted to copies.

-Keeping your projects small will make it easier to archive the entire project.

-To delete an entire project, you must first delete items in containers, then delete the container, then delete the project. Or your system administrator can use the *dmadmin* utility to delete the entire project and all relations.

Libraries:

-Use libraries to store your own parts and to share with others.

-Remember that a *Save* is required to make changes to a library permanent.

-Some data checking is performed when parts are checked into libraries, which adds to the level of protection you get when parts are stored in libraries rather than model files.

Tutorial: Working as a Team

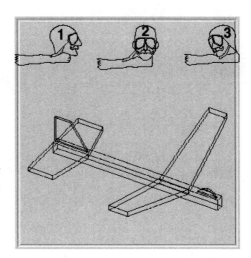

Libraries are the primary way that I-DEAS team members share information.

Before you start:

1. You should have a basic knowledge of creating and manipulating parts.

What you should learn:

1. How to use libraries in a collaborative team.

Tutorial Instructions.

Follow the instructions in the tutorials below.

Help, Help Library, Tutorials
Design Assemblies
Advanced Projects
Working as a Team

This tutorial will demonstrate how a team library can be used to share data. For this tutorial, you may either work with two friends, or work by yourself and switch between three model files to model a simple airplane.

Workshop 12A: Team Design

This workshop will let you apply what you have learned about part modeling, data management, and libraries to work with other users as a team to design a toy wooden car.

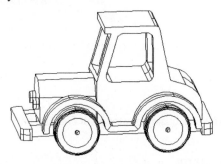

Before you start:

1. This workshop is designed to be performed with a team using a library to share data.

1. You should know how to use the basic functions of part modeling, data management, and libraries.

2. You must be working at a workstation that shares a common data installation with the other users who will work with you as a team.

After you're done, you should be able to:

1. Apply what you have learned to work as part of a design team.

Workshop Instructions:

Divide up into groups of two or three to model this wooden toy car as a team. Decide on a name for the project and library you will use, how you will position parts, what coordinate system, and what to name parts.

Start the I-DEAS software with a new model file. (Each team member will use their own model file.) Enter the Design application, Modeler task.

Workshop Instructions:

Start with the frame. One team member can create it and check it into the library where other team members can use it to build other parts in place around the frame.

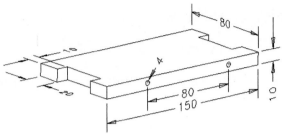

Dimensions are not given for other parts, so you can use your creativity. Decide between the members of your team who will build the body, engine, bumpers, fenders, lights, and wheels. Check these parts into the library as you create them.

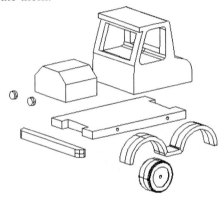

All team members should be able to get copies or references of parts out of the library to create the complete car on their workbench. Make multiple copies of common parts and position them where they belong.

Where parts are duplicated, you may either check multiple copies into the library (left and right versions), or just check one unique copy into the library. If you have already read Chapter 6, which covers assemblies, you may build an assembly using multiple instances of common parts rather than making copies in your model file. It would also be preferable to reference parts from the library, rather than making a copy, to preserve associativity, so that you can update the parts in your assembly from the library when changes are made.

When you are finished, either save and exit, or go on to the next workshop.

Workshop 12B:
Working With
Data Management

In previous workshops, you used I-DEAS Data Management when you named parts, put them away, got them out onto the workbench, and checked them in and out of libraries.

This workshop will investigate where data is stored and how you can find items by different attributes. The *Manage Bins* icon will be demonstrated, to show how to manage data in the I-DEAS software.

    Manage Bins

Before you start:

1. You should already know how to use the basic functions of the software for modeling.
2. This workshop assumes you have performed either the previous workshop or Workshop 2A, since these existing files will be used to demonstrate data management techniques.

After you're done, you should be able to:

1. Display metadata information and understand what it means.

Workshop Instructions:

Start the I-DEAS software with the model file from your last workshop or tutorial or the "PUMP" model file from Workshop 2A. Enter the Design application, Modeler task.

Select the *Manage Bins* icon to see what is in your model file.

   Manage Bins

One bin "Main" is created by default. When you get a part from a library, by default it's placed in a bin with the same name as the library. The *Manage Bins* form gives you a list of bins and the items they contain. I-DEAS Data Management (IDM) keeps track of the parts, assemblies, drawings, and other types of information called "items." Your model file is tracked as an item in IDM. It is also a container of other items. Your model file can be subdivided into "bins" to store parts.

You may get more information about these items by selecting them. For example, select one of your parts, then select *Details*, *Name*.

This will bring up the *Item Details* form, which gives more information on the name and part number of this part.

Notice also the *Parent Items*. The part is contained in the Main bin (or other). That bin is contained in the model file you are using.

Display the history of the part by pulling down to *History* on the selector at the upper left of the *Item Details* form.

The form will now show you information about when the part was created and modified. From this form, you can enter text information in the *Change History* field to describe why modifications were made. Other users can use this field to find out what is different about new versions before they actually update their references.

Cancel out of the *Item Details* form, but keep the *Manage Bins* form up.

The *Info* icon in the *Manage Bins* form will give you more specific information about items in your model file as known by the application you are working in. Pressing this button will generate information in the List window about the item you have selected. If you select an item not created by your current application, the Info icon may be grayed out.

The *Info* icon also gives specific information about referenced items as used by the application. If you request information on a part, you will be informed if that part is referenced in a drawing or an assembly. If you are using the model file in which you created a drawing, a finite element simulation, or an NC tool path, your part will be referenced by this other item in your model file.

Select different items in the *Manage Bins* form and then request information on them.

This is important for you to know, because the presence of a referenced item may prevent you from being able to delete a part in your model file. It could also be the case that you need to switch to the application that created the item to delete it, for example, if the referenced item is an NC Job, you will need to switch to Manufacturing to delete it.

In summary, the *Manage Bins* icon can give you information about the items stored in the bins in your model file.

Exit when you are finished.

Appendix A
Icon Summary

Function Keys

F1	Pan.
F2	Zoom.
F3	Rotate. 2D Rotation- Mouse at edge, 3D rotation- Mouse in center.
F4	View Snap (prior to I-DEAS 8, this function was Rotate Triad.

F5	Reset view.
F6	Bank select for other options F1-F5.
F7	Zoom All. Zoom and center rotation on selected entities, Zoom All if nothing selected.
F8	Reconsider.

F9	Deselect All.
F10	*(Not assigned)*
F11	Filter.
F12	Redisplay.

Print Screen	Create screen capture. (On Windows, use Alt-Print Screen to capture one window.)

Icon General Rules

An icon is highlighted when it is active. The icon highlighting, questions in the Prompt window, and change in shape of the cursor are types of feedback to tell you that the software is requesting input.

The triangle in the lower right corner indicates that more icons are below this icon. "Pull down" with the left mouse button ▣.

The rectangle in the lower right corner indicates that a sub-icon panel will be displayed with this icon.

The top and side bars over an icon indicate that this icon controls options.

The corner boxes imply selection, similar to the boxes around parts when they are selected.

A diamond shape around an icon indicates a manage activity, such as to store, get, and view the directory of stored items.

The square spiral shape means modify.

Arrows to the right indicate performing an action, such as to process the changes you have described to update the model.

Undo.

A large arrow to the right indicates performing an action that may take longer time to perform, such as to solve a simulation model.

Appearance. This "paint brush" icon is used to modify appearance attributes of parts and lines such as color. This "paint brush" shape appears in other icons along with other pictures, such as to modify the appearance of the workplane.

Get information, or generate a list in the list region. (Don't confuse this icon with getting information from the Help Library.)

Put something in a bin, catalog, or library. (Pointing up means "get.")

Delete. Remember that you can pick items before or after selecting the icon. To pick a list of more than one item, use the shift key while picking.

Icon Panel Layout

The icon panel is organized into the three icon sections and the top menu bar as shown below. The most important of these icons are summarized in the following pages.

| File Options   Help |
| Application   ▭ |
| Task   ▭ |

These pull-down menus were summarized Chapter 1. Notice that you can tell which application and task you are currently in by looking at the Application and Task pull-down menus.

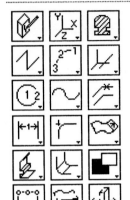

Task icons

The icons in this top section will change in each task of the current application. The icons shown here are for the Modeler task in the Design application. These will be explained more fully in the following pages, organized by application.

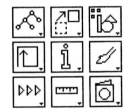

Application icons

The icons from this middle section generally stay the same in all tasks in an application. The icons shown here are the Design application icons.

View and Display icons

The icons in this section are I-DEAS-wide viewing and display commands. A summary of these icons starts on the next page.

View and Display Icons (I-DEAS-Wide)

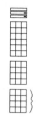

The icons in the lowest section of the icon panel appear in all I-DEAS applications. These control viewing direction and display type. Below is a summary of some of the important icons in this section of the icon panel.

Icon		Menu Equivalent	Description
	Redisplay	REDI	Redisplays the graphics in whatever display form is currently being used.

	Line	/DISPLAY_OPTIONS
		  LINE_HIDDEN_OPT
		    DISPLAY_TYPE
		      LINE
			Draw the display in line mode.

	Hidden
	  Hardware
	  Software
		/DISPLAY_OPTIONS
		  LINE_HIDDEN_OPT
		    DISPLAY_TYPE
		      HIDDEN
			Draws the display in normal hidden line mode. Hidden Hardware allows dynamic rotation of the hidden line display, if your display supports this feature.

	Shaded
	  Hardware
	  Software
		/DISPLAY_OPTIONS
		  SHADED_OPTIONS
			Draws the display in shaded image mode. (Other shaded options are available, depending on your display.)

	Zoom All	AU	Changes the view position so that entire display is sized to the viewport, or selected entities are centered.
	Zoom	ZM	Zooms in to an area of the display defined by dragging a box bounded by diagonal corners.
		/VIEW EYE	Key in your viewing direction (Eye Location).

	Center of
	  Rotation
		/VIEW
		  CENTER
			Pick a point to be the center of dynamic rotation.

	View
	  Workplane
		/VIEW
		  WORKPLANE
			Changes your view direction to be perpendicular to the workplane.

	One Viewport	/VIEW
		  DEFINE_LAYOUTS
		  MANAGE_LAYOUTS
			A stack of icons creates up to four standard viewports in the Graphics Window. Each viewport can have different display angles and options.

	Top View	/VIEW
		  EYE
		    0 1 0
			Changes the view direction to an XZ view, down the global Y axis. Similar icons set other standard orthogonal view directions.

	Isometric
	  View
		/VIEW
		  EYE
		    3,4,5
			Changes the view direction to an angled isometric view.

	Stop		Aborts a lengthy graphic display or calculation.

	Screen
	  Annotation
		/OPTIONS
		  SCREEN ANNOTATION
			Creates screen annotation for putting notes on the screen and your plots.

	Quick Print	/FILE QUICK PRINT	Print the workbench graphics. First, your administrator must define printers using *printAdmin*. Choose your default I-DEAS printer in *File, Print.*

Design Application

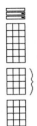

Design application icons.

Icon		Menu Equivalent	Description

	History Access	/CREATE
		    SPECIAL_TECHS
		        HISTORY TREE
			Displays the history tree. Select features, modify or rollback to a node.

	Move	/ORIENT MOVE	Moves parts, features, dimension text, and the workplane.
	Rotate	/ORIENT ROTATE	Rotate parts, features, or the workplane. Options include Slide on Screen and Measure.
	Align	/ORIENT ALIGN	Align parts or the workplane to other faces or reference planes.
	Drag	/ORIENT DRAG	Shape wireframe curves and sections dynamically by dragging points, curves, or dimensions. Drag arrow heads to change dimension in that direction. Drag text to reposition. Drag witness lines to reattach to an alternate location.
	Display Filter	/DISPLAY_OPTION DISPLAY_FILTERS	Turn entities on and off from the display.
	Modify	/MODIFY ENTITY	Modifies a part, or a dimension value.
Undo ◀◀◀	Undo	/UPDATE UNDO_LAST	Undo last Construction operation or variational geometry change.
	Info	/LIST INFO_ENTITIES	List information about a part, such as its topology and history.
	Appearance	/MODIFY APPEARANCE	Modifies color of parts; attributes of dimensions such as text size, arrows and add tolerance values. Change the standard from ISO to ANSI, etc. Use ⬚⬚■ Defaults to set defaults.
	Workplane Appearance	/WORKPLANE APPEARANCE	Set the limits of the 2D workplane. Grid snap on/off and grid display are an option on this form. (Grid was a separate icon prior to MS5.)
▷▷▷	Update	/UPDATE PROCESS_CHANGES	Re-process the history tree of a part to make changes that have been defined.
	Measure	/LIST MEASURE	Icon sub-panel to measure distances between points, lines, and planes.
Ixx	Properties	/LIST PROPERTIES	Compute mass properties of a part. Attach material to part.

Icon		Menu Equivalent	Description
	Parts...	/CATALOG PARTS...	Get standard parts from a catalog, including primitive shapes.
	Modify Catalogs	/CATALOG MODIFY CATALOGS	This icon brings up the Catalog sub-panel which contains the icons to "parameterize" parts and store them in catalogs. Also creates tables of configurations of parameterized parts.

	Parameters	/CATALOG
		   MODIFY CATALOGS
		     PARAMETERS
			Define which dimensions of a part or feature are to be variable parameters.

	Family Table	/CATALOG
		   MODIFY CATALOGS
		     FAMILY TABLE
			Create tables of the above parameters to define standard configurations of parameterized parts.

	Check In	/CATALOG
		   MODIFY CATALOGS
		     CHECK IN
			Check parameterized parts or features into a catalog.

	Get From Catalog	/CATALOG
		   MODIFY CATALOGS
		     GET FROM CAT
			Check out parameterized parts/features to modify their definition.

	Delete	/DELETE	Delete parts, lines, features, and constraints, etc.

	Put Away	/MANAGE
		   PUTAWAY
		   GET
			Put part away in a bin, remove from workbench. Opposite of Get.

	Name Parts	/MANAGE NAME	Give parts a permanent name, but leave them on the workbench.
	Manage Bins	/MANAGE MANAGE_BINS	Manage the model file bins where items are stored. Rename and copy parts, create new bins. In I-DEAS 8, this icon is removed. Use *File*, *Manage* to manage model files.
	Check In	/MANAGE CHECK IN	Check parts, assemblies, and drawings into a library.
	Get From Library	/MANAGE GET FROM LIBRARY	Get parts, assemblies, and drawings from a library. Library status can be Check Out to modify, Reference, or Copy (with or without notification).
	Update From Library	/MANAGE LIBRARIES	Update parts or assemblies to the latest version. (Do not Get From Library again to update.)
	Manage Items	/MANAGE ITEMS	Manage all items in a data installation. Delete Files from within the software. Initially lists all the model files in the current project. Manage Projects, Manage Libraries, and Manage Catalogs icons are similar, but initially list only those items.

Modeler Task

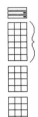

Modeler task icons.

These icons are organized around the modeling process:
1. Select a workplane, (row 1)
2. Sketch and constrain, (rows 2-4)
3. Create feature (rows 5,6)

Icon		Menu Equivalent	Description
	Create Part	/WORKPLANE NEW PART	Create and name a new part starting with a datum coordinate system.
	Sketch in Place	/WORKPLANE ATTACH	Sketch wireframe geometry on a face of a part to cutout, intersect, protrude. An option is to sketch on path.
	Sketch on Workplane		Revert to sketching on global workplane.

	Coordinate Systems	/CREATE
		    REFERENCE GEOM
		        COORDINATE SYS
			Create local coordinate systems that can be used to align features or other parts. This stack also creates reference points and lines. Create a coordinate system and name it as a part to create a "BORN" part.

	Reference Planes	/CREATE
		    REFERENCE GEOM
		        PLANE
		        POINT
		        LINE
		        CURVE
		        PLANE
			Create reference planes to create or align features. Sketch in Place works with reference planes like any other face of the part. This stack also contains icons to create reference lines and points. Reference curves can be used when curves are used in more than one feature, so they will not be "absorbed" when used.

	Build Section	/CREATE SECTION	Create a section from wireframe curves. (You can skip this step using the *Extrude/Revolve* commands.)

	Extract	/CREATE
		    SPECIAL TECHS
		        EXTRACT
			Extract a feature as a part, extract the wireframe definition of a feature, or extract a wireframe boundary of a part.

	Attach	/CREATE
		    SPECIAL TECHS
		        ATTACH
			Attach wireframe geometry to a part. To align this geometry with a particular face, use the *Align* command. To associatively relate the geometry to the face, add at least one dimension to tie the new geometry to the edge of the face.

	Cross Section	/CREATE
		    SPECIAL TECHS
		        CROSS SECTION
			Section a part, surface, or curve.

Sketching and Constraining

Icon		Menu Equivalent	Description
	Polylines	/CREATE POLYLINES	Creates Polylines end to end on workplane or face of part.
	Lines	/CREATE SINGLE_LINES	Creates single lines, by start and end point.

	Rectangle By 2 Corners	/CREATE
		   RECTANGLE
		   RECT_BY_2_CORNERS
			Creates rectangles on workplane or face of part. This stack also contains other methods to create rectangles such as by three corners.

	Polygon	/CREATE POLYGON	Create constrained wireframe polygons. (I-DEAS 8 or later.)
	Points	/CREATE POINTS	Create points.

	Three Points On	/CREATE
		   ARC
		   THREE PTS ON
			Creates arcs through any three points. Use this command to create an arc, then constrain tangencies, etc.

	Center Start End	/CREATE
		   ARC
		   CENTER_START_END
			Creates arcs counter-clockwise about center, from point 2 to point 3.

	3D Lines	/CREATE
		   3D METHODS
		   POINT, LINE
		   CIRCLE, ARC
		   SPLINE, etc
			This stack of icons creates 3D curves and points not on a 2D workplane.

Icon		Menu Equivalent	Description

	Center Edge	/CREATE
		   CIRCLE
		   CENTER_EDGE
			Creates 360 degree circles.
			Other circle creation methods also in this stack. Use *Circle Tangent to 3* to create elongated slots.

	Splines	/CREATE SPLINE	Creates splines. Use *Function Spline* to create a spline from an equation
	Ellipse by 2 Corners		Create ellipses.

	Offset	/MODIFY
		   WIREFRAME
		   OFFSET
			Offset sections or single curves to create a new section.
			Other icons in this stack create curves on surfaces such as by *Through Points*, *Silhouette Curves* on surfaces, and *Iso Curves* at constant S,T values.

	Project Curve	/CURVE ON SURFACE PROJECT CURVE	Project a curve onto a surface.
	Surface Intersection	/CURVE ON SURFACE SURF INTERSECTION	Create a spline at the intersection of two surfaces.

Sketching and Constraining (Cont.)

Icon		Menu Equivalent	Description
	Dimension	/CONSTRAIN DIMENSION	Create linear, radial, or angular dimensional constraints. While dimensioning, the Dimension Advisor uses these colors: Green = unconstrained geometry. Red = partially constrained geometry. Blue = fully constrained geometry.
	Part Equations		Add equations to dimension networks. Remember to use		to enter units.
	Constrain & Dimension	/CONSTRAIN	Create variational geometry constraints on wireframe curves. Displays sub-panel shown below.
	Parallel	/CONSTRAIN PARALLEL	Constrain lines to be parallel.
	Perpendicular	/CONSTRAIN PERPENDICULAR	Constrain lines to be perpendicular.
	Tangent	/CONSTRAIN TANGENT	Constrain lines and arcs to remain tangent.
	Coincident & Colinear	/CONSTRAIN COINCIDENT COLLINEAR	Makes lines or points coincident. Use to keep circles centered with the same center points.
	Linear	/CONSTRAIN LINEAR_DIM	Create linear dimension constraints. This will hold a dimension constant until it is modified or "dragged."
	Angular	/CONSTRAIN ANGULAR_DIM	Create angular dimension constraints.
	Radial	/CONSTRAIN RADIAL_DIM	Create radial dimension constraints.
	Anchor	/CONSTRAIN GROUND ANCHOR FIX_ANGLE HORIZONTAL VERTICAL	Ground points, and fix lines to be horizontal, vertical, or to remain at the specified angle.
	Fix Angle		Grounding points and lines that you don't want to move should be one of the first things you do to your wireframe geometry.
	Horizontal Ground		
	Vertical Ground		The *Lock* and *Fuse* commands are used to constrain assembly instances relative to one another.
	Show Free	/CONSTRAIN SHOW_CONSTRAINTS SHOW_FREE SHOW_CAUSES SHOW_EFFECTS SHOW_CONSTRAINT	Use these icons to check what freedoms are constrained or left free.
	Show Causes		Red = Partially constrained geometry. Blue = Fully constrained geometry.
	Show Effects		Red arrows show unconstrained freedoms.
	Show Constraints		This stack also contains *Animate* to animate dimension changes.
	Auto Constrain	/CONSTRAIN AUTO_CONSTRAIN	Create constraints and dimensions automatically within specified tolerances set using mouse button 3.

Sketching and Constraining (Cont.)

Icon	Menu Equivalent	Description
Shape Design	/SHAPE_DESIGN	Variational Shape Design. Advanced curve/surface shaping.
Shape		The first row of icons performs direct curve shaping. The second row applies constraints to curves. The third and fourth row perform energy-based and curvature shaping. These commands iteratively shape curves based on constraints, forces (push, twist, magnet), and the "prototype geometry" applied to the curve.
Drag Point	/SHAPE_DESIGN MODIFY MOVE POINT ROTATE POINT	Shape curves and surfaces. To move normal to plane, use the option: ☐☐☐ *Move_Normal.* Other commands in this stack are *Drag Tangency* and *Flip Tangency*.
Flip Tangent	/SHAPE_DESIGN MODIFY FLIP TANGENT	Flip the tangent around where two curves intersect.
Drag Tangent Influence	/SHAPE_DESIGN MODIFY TANGENT INFLUENCE	Dynamically set the tangent influence "strength."
Expand/ Contract	/SHAPE_DESIGN MODIFY EXPAND	Pick a point on a curve, and then "push" it away from the curve center based on the cursor position.
Ground End	/SHAPE_DESIGN CONSTRAINTS GROUND POSITION	Ground the end location of curves. similar commands to ground the tangent and curvature of curve ends.
Coincident Points	/SHAPE_DESIGN CONSTRAINTS	The icons in this stack constrain the intersection between two curves to be coincident, tangent, or to match curvature.
Push Twist	/SHAPE_DESIGN CONSTRAINTS PUSH TWIST	The icons in this stack *Push* and *Twist* with the equivalent of point forces or magnets to shape curves using "energy-based" shaping.
Prototype Geometry	/SHAPE_DESIGN CONSTRAINTS PROTOTYPE GEOM	Prototype geometry is the natural free state the curve will tend toward in the absence of other forces, much as a stretched rubber band tends toward a straight line.
Function	/SHAPE_DESIGN FUNCTION	Use Function, Create, From Curve to create XY functions of curvature, etc. from existing curves. Edit functions to get precise control on curvature. Use functions for prototype geometry to shape other curves "like" this one.
Full Refine	/SHAPE_DESIGN REFINE	Add more control point degrees of freedom to a curve.

Tolerance Analysis

Icon		Menu Equivalent	Description
	Tolerance Analysis	/ANALYZE GEOMETRY TOLERANCE ANAL	Display the Tolerance Analysis sub-panel of icons.
	Specify Tolerances	/ANALYZE GEOMETRY TOLERNCE ANAL SPECIFY TOLERAN	Apply manufacturing tolerance bounds to driving dimensions, and allowable tolerance bounds on reference dimensions.
	Default Tolerances	/ANALYZE GEOMETRY TOLERNCE ANAL DEFAULT TOLER	Specify default tolerances for geometric constraints.
	Verify	/ANALYZE GEOMETRY TOLERNCE ANAL VERIFY	Check the constraint network to make sure it is a valid tolerance model.
	Dimension To Analyze	/ANALYZE GEOMETRY TOLERNCE ANAL ENTITY TO ANALYZ	Pick the (reference) dimension to be analyzed.
	Standard Analysis	/ANALYZE GEOMETRY TOLERNCE ANAL ANALYSIS	Execute the analysis using the standard method, comprehensive method, or to request a sensitivity analysis.

Model Views and PMI

Icon		Menu Equivalent	Description
	Annotation	/ANNOTATE	Depending on your license, this icon displays the complete Master Notation panel including model views, or the basic annotation icon panel. Create Product and Manufacturing Information (PMI) notation.
	Model Views	/ANNOTATE MODEL VIEWS	Create and activate model views. (I-DEAS 8 or later.)
	Notes	/ANNOTATE NOTE	Create notes on 3D parts and assemblies.
	Surface Finish	/ANNOTATE SURFACE FINISH	Create surface finish symbols on 3D parts and assemblies..
	Feature Control Frame	/ANNOTATE FEAT CONTR FRAME	Create GD&T feature control frames.
	Datum Feature Symbol	/ANNOTATE DATUM FEAT SYMB	Create datum reference feature symbol.
	World Wide Web	/LIST WORLD WIDE WEB	Link to URL placed in notes.

Wireframe/Surface Tools

Icon		Menu Equivalent	Description
	Fillet	/MODIFY WIREFRAME FILLET	Wireframe fillet. (Not the same as fillet of surfaces.) This stack also contains *Make Corner* to form an intersecting corner between two curves.
	Trim/Extend	/MODIFY WIREFRAME TRIM/EXTEND	Trim and extend lines. (It is not necessary to trim curves to make sections.) See also icons to divide curves and merge curves.
	Trim at Curve	/MODIFY TRIM SURFACE	Trim Surfaces to new bounding edges. Icons in this stack modify surfaces and result in new or modified surfaces.
	Extend Surfaces	/MODIFY EXTEND SURFACES	Extend Surfaces beyond existing edges.
	Interpolate Surfaces	/CONSTRUCT INTERPOLATE SURF	Create an interpolated surface between two surfaces.
	Merge Surfaces	/MODIFY MERGE SURFACES	Merge two surfaces to create a new (blended) surface.
	Divide Edge	/MODIFY DIVIDE EDGE	Divide the edges of surfaces. This stack also includes *Unwrap Surfaces*, (I-DEAS 8 or later.)

Creating Features

Icon		Menu Equivalent	Description
	Extrude	/CREATE EXTRUDE REVOLVE	Extrude a section to make a new part or to cut/join with an existing part. This stack also contains *Revolve*.
	Sweep	/CREATE SWEEP	Sweep sections or surfaces along a path curve. The path may be the edge of a part. The path slope must be continuous. Variational Sweep allows the section to change along the path, driven by constraints.
	Loft	/CREATE LOFT	Loft between sections or surfaces to create parts.
	Surface By Boundary	/CREATE SURFACE_BY_ BOUNDARY	Create a surface from the bounding curves.
	Mesh of Curves	/CREATE MESH_OF_CURVES	Create surfaces from curves in 2 directions.
	Variational Surface Feature	/CREATE VARIATIONAL_ SURF_FEAT	Create free-form history-based features on surfaces. (I-DEAS 8 or later.)

Creating Features (Cont.)

Icon	Menu Equivalent	Description
Fillet	/CONSTRUCT FILLET	Fillet/round edges of parts. This command works with edges, but you can pick surfaces or vertices to select adjacent edges.
Chamfer	/CONSTRUCT CHAMFER	Chamfer edges of parts.
Shell	/CONSTRUCT SHELL	Create a thin-walled part from a surface or from the surfaces of a solid part. Thicken a surface into a solid.
Draft	/CONSTRUCT DRAFT	Change the draft angle of part faces.

Midsurface	/CONSTRUCT
	  SURF ABSTRACTIONS
	    MIDSURFACE
		Create a new surface at the midsurface of a part by pairing opposing faces manually or automatically. Midsurface is especially useful for finite element modeling.
		Options control the size of small details to ignore and other parameters. Other surfaces other than the midsurface may be deleted.

Cut	/CONSTRUCT
	  CUT
	  JOIN
	  INTERSECT
		Cut one part from another.
		Also in this stack of icons are *Join* and *Intersect*. The *Join* command can be used to join wireframe parts if they are named first.

Turn Relations On	/CONSTRUCT RELATIONS_SW	To include associative design intent relations on construct operations, this switch must be on when construct or reflect operations are performed.
Plane Cut	/CONSTRUCT PLANE_CUT	Cut a part with a plane.
Partition	/PARTITION	Divide a part into multiple volumes. Used to create different regions for finite element modeling. A similar icon is *Split Surface*, which does not create an internal surface to create volumes.
Stitch Surface	/MODIFY STITCH SURFACE	Join surfaces along common edges.

Creating Features (Cont.)

Icon		Menu Equivalent	Description
	Reflect	/MODIFY REFLECT	Reflect parts or sections
	Scale	/MODIFY SCALE	Scale parts larger or smaller by a scale factor, or nonuniformly by scale factors in x, y, and z directions.

	Rectangular Pattern
	
	Circular Pattern
		/CONSTRUCT
		    CIRCULAR PATTERN
		    RECT PATTERN
			Create a rectangular pattern of a part to create a feature, or to make a rectangular pattern from an existing feature. Similar command to create a circular pattern.

Diagnostics and Clean-up

Icon		Menu Equivalent	Description
	Show Free Edges	/ANALYZE GEOMETRY SHOW FREE EDGES	Show edges of surfaces not connected to other surfaces. Also, use Exploded to show each surface.
	Surface Cleanup Auto Trim	/MODIFY AUTO TRIM	Defines closed loops, merges surfaces together, and creates new surfaces inside the loop boundary. Use this command for cleaning up geometry imported from other CAD systems, or preparing surfaces for finite element meshing.

	Clean Coincident Points	/MODIFY
		    CLEAN
		        COINC POINTS
			Use these icons to "clean" geometry brought in from other CAD systems to clean up duplicate points and curves

	Material Side	/MODIFY
		    SPECIAL TECHS
		        MATERIAL SIDE
			Define which side of a surface acts like it has material during construction operations. Normally closed parts have material to the inside of all bounding surfaces. In open-part modeling, you may need to set this.

	Surface Quality	/ANALYZE GEOMETRY SURFACE QUALITY	This stack of icons contains tools to analyze surface curvature, tangent and normal.
	Diagnose Part	/ANALYZE GEOMETRY DIAGNOSE PART	Check the validity of a solid part. (I-DEAS 8 or later.)

Mold Design

Icon		Menu Equivalent	Description
	Die Lock Check	/ANALYZE GEOMETRY DIE LOCK CHECK	Check draft angles for moldability. (See also the *VGX Core Cavity Die Lock Check* command.)
	VGX Core Cavity		Icon panel with commands to split a mold block into core and cavity (A/B) sides. (I-DEAS 8 or later.)
	VGX Mold Base		Icon panel with commands to create a mold base assembly and add mold components.

Sheet Metal

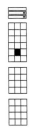

Sheet Metal icons. (In the Modeler task.)

Icon		Command Equivalent	Description
	Sheet Metal..	/SHEET METAL	Bring up the sub-panel below to perform sheet metal operations.
	Build Panel	/SHEET METAL CREATE PANEL FROM WIREFR	Build panels from wireframe geometry.
	Sheet Metal	/SHEET METAL CREATE SHEET METAL	Create a sheet metal part from panels or from faces of a part.
	Create Bend	/SHEET METAL CREATE BEND	Create Bends, between panels, assigning radius value and K factor.
	Unfold Panels	/SHEET METAL UNFOLD PANEL	Unfold the sheet metal part.
	Fold Panels	/SHEET METAL FOLD PANEL	Fold the sheet metal part.
	Ground Panel	/SHEET METAL MODIFY GROUND PANEL	Define one panel to be the grounded panel that does not move when the part is folded.
	Punch Hole	/SHEET METAL PUNCH	Punch holes in the sheet metal part.
	Appearance	/MODIFY APPEARANCE	Pick this command, hold the right mouse button, select *Defaults*, then *Sheet Metal*, to set default bend angle and radius.
	Modify	/MODIFY ENTITY	Modify bend parameters after creating bends. Use *History Access* to pick the sheet metal solid data to fold and unfold the part.
	Materials	/MATERIALS	Assign materials. Create material types as "Sheet Metal K Factor." Assign thickness, density, and K factor.
	Shell	/CONSTRUCT SHELL	Thicken into a solid part.

Assembly Task

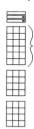   Assembly task icons.

Icon		Menu Equivalent	Description
	Hierarchy	/ASSEMBLE HIERARCHY	Define assembly hierarchy. Add parent to start a new Assembly.
	Add to Assembly	/ASSEMBLE ADD TO ASSEMBLY	Add instances of parts to an assembly.
	Remove from Assembly	/ASSEMBLE REMOVE INSTANCE	Delete instances from an assembly.

	Replace Instance	/ASSEMBLE
		   REPLACE
		   INSTANCE
			Replace instances in an assembly.

	Duplicate Instance	/ASSEMBLE
		   DUPLICATE
		   INSTAN
			Duplicate instances in an assembly.

	Make Unique	/ASSEMBLE MAKE UNIQUE	Make a copy of a part instanced in an Assembly, so that the instance is unique.

	Suppress	/ATTRIBUTES
		   SUPPRESS
		   UNSUPPRESS
			Suppress instances from the display of the assembly.

	Constrain & Dimension	/CONSTRAIN	Create dimensional relationships between instances. You should Lock at least one instance to the parent assembly, then constrain each instance in a logical sequence, constraining all 6 DOF.
	Browse Relations	/LIST RELATIONS BROWSER	Display relations between instances, including VGX constraints and associative copy relations.
	Manage Configuration	/CONFIGURATIONS MANAGE	Manage configurations, store for use in animation, interference checking.
2,4,8	Create Sequence or Manage Sequences	/ANIMATION CREATE SEQUENCE	Create a sequence of configurations for animation. (In I-DEAS 8 or later, *Create Sequence* icon is on the *Manage Sequences* form.)
	Animate Hardware or Animate Software	/ANIMATION ANIMATE	Display sequences of configurations in animation.
	Interference	/LIST INTERFERENCES	Check for interferences between instances.

Mechanism Design Task

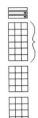

Mechanism Design task icons. Use this task to analyze mechanism kinematic motion.

In I-DEAS 8 or later, use Mechanism Simulation task in the Simulation application to analyze forces.

Icon		Menu Equivalent	Description
	Mechanism Info	/LIST MECHANISM DATA	List information on mechanism joints and markers.
	Create Functions	/FUNCTION CREATE	Create functions for force or motion input. Modify or Delete functions. *(Make sure all menus are turned on.)*
	Graphing Graph XY		Plot functions created by the user or functions created by the solve.
	Create Motion	/LOADS MOTION	Create driving motions of the mechanism by applying a function to a joint variable.
	Create Force	/LOADS FORCE	Create Forces. (This icon is only in the Mechanism Simulation task.)
	Create Gravity	/LOADS GRAVITY	Create Gravity Loads.
	Initial Conditions	/LOADS INIT CONDITION	Supply initial motion conditions.

	Constraints
	
	Revolute Joint
		/JOINT/CONSTRAINTS
		    JOINTS
		        REVOLUTE
			Create joint relationships between instances. This stack contains several types of kinematic joints.

	Translation-al Joint	/JOINT/CONSTRAINTS
		    JOINTS
		        REVOLUTE
			Create a translational joint with one sliding DOF.

	Gear or Coupler	/JOINT/CONSTRAINTS MULTI JOINTS	Create gear joints.

	Manage Joints	/JOINT/CONSTRAINTS
		    JOINTS
		        MANAGE
			List information about joints, delete and rename joints.

	Spring Damper	/JOINT/CONSTRAINTS SPRING/DAMPER	This stack of icons creates spring/damper relationships.
	Create Ground	/CONSTRAIN GROUND	Ground an instance in the mechanism.
	Create Rigid Body	/CONSTRAIN RIGIDIFY ASSEM	Make a subassembly act like one rigid body.
	Solve Setup, Solve	/SOLUTIONS	Create and manage load cases, and solve the mechanism.

Drafting Setup Task (Design Application)

▶▶▶ This task has been removed in I-DEAS 8 or later. This page only applies to older versions of the software.

Icon		Menu Equivalent	Description
	Create Layout	/CREATE LAYOUT	Create a new drawing from a part or assembly.

	Front	/CREATE VIEW
		   STANDARD
		     FRONT
		     TOP
		     ISO
		     (etc.)
			Add standard views to the drawing.

	Section	/CREATE VIEW
		   SECTION
		   DETAIL
		   AUXILIARY
		   USER DEFINED
			Create views derived from existing views on the drawing.

	Activate View	/ACTIVATE VIEW	Pick the view to work in.
	Move	/MODIFY VIEW MOVE	Move views on the page.
	Scale Views	/MODIFY VIEW SCALE VIEWS	Change the scale factor of views.

	Linear	/DETAILS
		   DIMENSION
		     LINEAR
		     ANGULAR
		     RADIAL
		     HORIZONTAL
		     VERTICAL
		     (etc.)
		   NOTE
		   BALLOON
		   CROSSHATCH
			Create drawing annotation.

	View Borders	/OPTIONS
		   DRAWING SETTINGS
		     VIEW BORDERS
			Turns the display of view borders on and off.

	Detailing	/DETAILING	Bring the drawing into the Drafting application for further detailing.
	Export	/EXPORT	Export the drawing file in various formats to use in other drafting packages.

Drafting Task

 These pages describe I-DEAS 8 or higher. In older versions of the software, Drafting was an application, not a Design task. For these versions, most commands are similar. A major difference is how drawings of parts and assemblies are created.

 The top menu bar contains the menus *File*, *Options*, and *Help*. Use the online Help library to get more detailed information.

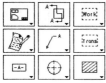 This section of the icon panel contains icons for creating a drawing, creating views, and adding annotation such as notes and GD&T frames. Start with the icon in the upper left to create a drawing.

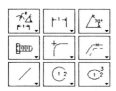 The second section contains icons for creating drawing entities including lines and dimensions.

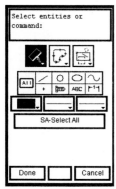 The middle section contains the Command Options Area (COA), which changes to show options available in each command. Selection Options are shown when no command is active. Location Options are shown for curve creation commands.

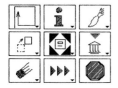 This section of the icon panel contains commands for editing, listing, modifying, and deleting.

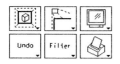 The very bottom section contains view control for zooming and redrawing graphics. The lower left corner contains the *Undo* and *Redo* commands.

Drafting Task (Cont.)

Icon		Menu Equivalent	Description
	Create Drawing	DQ	Create a new drawing from a part, assembly, or from scratch. Use *View Drawing* in this stack to view old drawings without modifying them.
	Section View	VX	Create section views of parts and assemblies. This stack also contains *Detail View* and *Aux View*.
	Workview	W	You may also just click in a view to make it the active view. Use other icons in this stack to add views to existing drawings and modify view borders. Use *View Properties* to modify view hidden line processing.
	Note	NN	Add text notes to the drawing..
	Label	DL	Add label notes and balloon callouts.
	Transfer	VT	Transfer geometry from one view to another. also in this stack is *Move Views* to move the whole view, and *Delete Views*.

Simulation Application

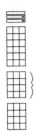

Simulation application icons.

Icon			Menu Equivalent	Description
	`⌧`	`Name Parts..`	`/MANAGE` `NAME...`	Remember that parts must be Named (or Put Away which also gives them a name) before they can be used for meshing.
		`Create` `FE Model`	`/MANAGE` `CREATE FE MODEL`	Creating an FE model associates the model with the part. All FE models must be associated with a part, even if the part is empty.
		`Manage` `FE Model`	`/MANAGE` `MANAGE FE MODEL`	Create "Design Studies," where the same FE model can have multiple solutions with different conditions. (I-DEAS 8 or later.)
		`Create FEM` `From ASM`	`/MANAGE` `MANAGE FE MODEL`	Create an FE model associated with an assembly. (I-DEAS 8 or later.)
		`Fem Groups`	`/GROUP`	Create groups of FE entities. Group icon brings up the Group icon sub-panel.

Modeler Task for Simulation

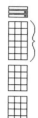

Modeler task icons.

For Modeler task icons, see the Design application. The icons listed below describe specific usage of these icons for the Simulation application.

Icon		Menu Equivalent	Description
	Build Section	/CREATE SECTION	Create a section from wireframe curves. A section is only a boundary definition, it does not contain an interior surface for meshing.
	Cross Section	/CREATE SECTION	Create a planar, bounded surface by cutting a part with a plane. Use this command to create a surface that can be used for 2D meshing on an arbitrary cross section through a part.
	Trim at Curve	/MODIFY TRIM SURFACE	Trim surfaces to new bounding edges. (Useful for creating boundary conditions on a limited surface area.) Icons in this stack modify surfaces and result in new or modified surfaces.
	Extend Surfaces	/MODIFY EXTEND SURFACES	Extend surfaces beyond existing edges.
	Interpolate Surfaces	/CONSTRUCT INTERPOLATE SURF	Create an interpolated surface between two surfaces.
	Merge Surfaces	/MODIFY MERGE SURFACES	Merge two surfaces to create a new (blended) surface.
	Divide Edge	/MODIFY DIVIDE EDGE	Divide the edges of surfaces.

Modeler Task for Simulation (Cont.)

Icon	Menu Equivalent	Description
Surface by Boundary	/CREATE SURF BY BOUNDARY	Create surfaces from a set of curves on the boundary. Use this command to create a surface to use to mesh a 2D finite element model from a wireframe section.
Midsurface	/CONSTRUCT SURF ABSTRACT MIDSURFACE	Create a new surface at the midsurface of a part by pairing opposing faces. Midsurface is especially useful for finite element modeling using thin-shell elements.
Midsurface Attributes	MIDSURFACE ATTR	Midsurface attributes control the size of small details to ignore and other parameters.
Show Midsurface Only	/CONSTRUCT SURF ABSTRACTIONS SHOW MID ONLY	Hide other surfaces other than the midsurface.
Delete Midsurface	/CONSTRUCT SURF ABSTRACT DELETE MIDSURF	Delete the midsurface.

Icon	Menu Equivalent	Description
Partition	/PARTITION	Divide a part into multiple volumes using another part as a cutting tool. Use this command to mesh volumes that touch, but have different material properties. Each volume can be assigned different attributes for meshing. Use this command also to divide surfaces for local surface boundary conditions.
Plane Cut	/CONSTRUCT PLANE_CUT	Cut a part with a plane. Similar to the *Cross Section* command above, but the result is the positive or negative half of the part, not just the cross section. This command may also be useful for meshing a 2D model on an arbitrary cut section of a part.
Stitch Surface	/MODIFY STITCH SURFACE	Join surfaces along common edges.

Icon	Menu Equivalent	Description
Surface Cleanup Automatic Trim	/MODIFY AUTO TRIM	Defines closed loops, merges surfaces together, and creates new surfaces inside the loop boundary. Use this command for cleaning up geometry imported from other CAD systems, or preparing surfaces for finite element meshing.
Clean Coincident Points	/MODIFY CLEAN COINC POINTS	Use these icons to "clean" geometry brought in from other CAD systems to clean up duplicate points and curves

Meshing Task

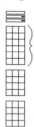 } Meshing task icons.

Icon		Command Equivalent	Description
	Define Shell Mesh	/DEFINE SHELL MESH	Set meshing parameters on surfaces.
	Define Solid Mesh	/DEFINE SOLID MESH	Set meshing parameters on part volumes.
	Define Beam Mesh	/DEFINE BEAM MESH	Set meshing parameters for beams on part edges or wireframe curves.
	Define Free Local	/DEFINE FREE LOCAL LENGTH	Define local densities for free meshing.
	Modify Mesh Preview	/PREVIEW MESH	Preview the size of elements and calculate element quality.
	Shell Mesh	/GENERATE SHELL MESH	Generate thin-shell elements on a surface.
	Solid Mesh	/GENERATE SOLID MESH	Generate solid elements on part volumes.
	Beam Mesh	/GENERATE BEAM MESH	Generate beam elements on part edges or wireframe curves.
	Mesh on Part	/GENERATE PART	Generate elements on surfaces and volumes of a part. (Part may contain "non-manifold" topology.)
	Quality Checks	/CHECK MESH DISTORTION & STRETCH	This command will help find distorted or stretched elements which may lead to inaccurate results.
	Auto Settings	/CHECK MESH AUTO SETTINGS	Use this icon to identify the checks that will be performed on the mesh when it is generated, and set values to be used.
	Mesh Delete	/UNDO MESH	Delete the entire mesh and start over.
	Tetra Fix	/CHECK MESH TETRA FIX	Attempt to fix distorted tetrahedral elements.

Meshing Task (Cont.)

Icon		Menu Equivalent	Description
	Node	/NODE CREATE...	Create nodes manually.
	Modify (Node)	/NODE MODIFY	Modify node coordinates or color.
	Copy	/NODE COPY	Copy a pattern of nodes to generate new nodes.
	Between Nodes	/NODE BETWEEN NODES	Create new nodes between two sets of nodes.
	Element	/ELEMENT CREATE...	Create elements manually.
	Modify (Element)	/ELEMENT MODIFY	Modify element properties.

	Copy to
	Existing
	Nodes
		/ELEMENT
		  COPY TO
		  EXISTING NODES
			Copy elements using a pattern of nodes already created.

	Reflect	/ELEMENT REFLECT	Reflect nodes and elements.
	Materials	/MATERIALS...	Create and modify materials.
	Physical Properties	/PHYSICAL PROP	Create and modify physical properties, such as thickness of shell elements.
	(No icon, menu only.)	/NODE BANDWIDTH MANAGE	Commands to minimize the bandwidth of the model. (Only needed for external solvers.)
	(No icon, menu only.)	/ELEMENT WAVEFRONT	Commands to minimize the element wavefront of the model. (Only needed for external solvers.)

Boundary Conditions Task

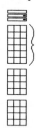

Boundary Conditions task icons.

Icon		Command Equivalent	Description
	Linear Statics	/ANALYSIS TYPE	Select linear analysis type. This will cause unnecessary icons to be grayed out.
	Normal Mode Dynamics		Select dynamics analysis type.
	Force	/CREATE FORCE	Apply forces to part vertices, edges, and surfaces or nodes. Applying forces at point locations on edges and surfaces will generate a node at that location during meshing.
	Pressure	/CREATE PRESSURE	Apply pressures to surfaces of parts or element faces.
	Displacement Restraint	/CREATE RESTRAINT	Apply displacement restraints on part geometry or nodes.
	Boundary Conditions	/BOUNDARY CONDITION	Group Restraint Sets, Load Sets, etc., into a Boundary Condition Set.
	Sets	/SETS	Create new sets and select which sets are "current" (active) of loads, restraints, DOF sets, etc.

Model Solution Task

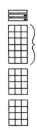

Model Solution task icons.

Use this task to solve for statics or dynamics, then review the results in the Post Processing task or using the *Visualizer* command.

Icon		Command Equivalent	Description
	Solution Set	/SOLUTION SET	Create a Solution Set containing the type of solution, the Boundary Condition Set, Output Selections, and other options, such as Verification Only.
	Manage Solve	/SOLVE	Set solve options such as Interactive or Batch, location of Hypermatrix file, name of output list (log) file.
	Solve	/SOLVE	Solve.
	Report Errors/ Warnings	/REPORT SOLUTION ERROR	Review solution warnings and errors.

Response Analysis Task

I-DEAS 8 or later - For older versions, see the online help manuals

Icon		Command Equivalent	Description
	Function Tools	/FUNCTION OPERATIONS CREATE	Create time and frequency functions for dynamic input events. Graph input and response functions.
	Create Event	/CREATE EVENT	Create an event defining static or dynamic event type. For transient dynamics, you will also define the modal results from a previous normal modes analysis. Add the input functions to the event.
	Manage Events	/MANAGE EVENTS	Manage the naming and storage of events.
	Nodal Functions Elemental Functions	/FUNCTION RESULTS NODAL ELEMENTAL	Compute response function results at specific DOF from a dynamic model.
	Evaluate Results	/RESPONSE EVALUATION	Compute result sets to display on the entire model.
	Strength Evaluation	/STRENGTH EVALUATION	Calculate a result set of the strength. Display as a contour plot.

Durability Task

I-DEAS 8 or later - For older versions, see the online help manuals.

Icon		Command Equivalent	Description
	Function Tools	/FUNCTION OPERATIONS CREATE	Create input functions to scale static results or to define duty cycles.
	Create Event	/CREATE EVENT	Create an event defined with a static event type to evaluate fatigue to repeated static loads. Use a transient event type to compute fatigue based on dynamic results.
	Manage Events	/MANAGE EVENTS	Manage the naming and storage of events.
	Strength Evaluation	/STRENGTH EVALUATION	Calculate a result set of the strength. Display as a contour plot.
	Evaluate Fatigue Safety	/FATIGUE SAFETY EVAL	Calculate a result set of fatigue safety factor. Display as a contour plot.
	Evaluate Fatigue	/FATIGUE EVALUATION	Calculate the fatigue damage to a number of occurrences of the event.
	Fatigue Tools		This icon displays an icon panel to calculate and display fatigue results using histogram-counting tools. Use icons on this panel to define fatigue material properties.

Optimization Task

Optimization task icons.

Icon		Description
	Design	Create a "Design", which describes the optimization or parameter study. Note that once a design is created, the FE model used in it can't be modified unless the design is deleted.
	Solution Set	Create a solution set the same way as in the Model Solution task. This set defines the output selections.
	Step Control	Define the number of positive and negative steps from current value for design parameter studies.
	Iteration Control	Define the maximum number of iterations and factors that control how much the parameters are allowed to change between each iteration, and a convergence tolerance to determine when to end.
	Design Parameters	Define each design parameter. For geometry-based design parameters select the geometry controlled by the design parameter. For parameter studies, you should only define one design parameter, since multiple design parameters will be modified together at each step, not independently as in optimization.
	Stress Limits	Define optimization limits on stress. Other limits can be set for mass, displacement, and natural frequency.
	Stress Monitor	Define regions in which to graph stress as a function of design parameters.
	Design Goal	Used for optimization. The default design goal is to minimize total mass.
	Displacement Monitor	Define locations to graph displacement.
	Mass Monitor	Define regions or volumes to graph mass.
	Sensitivities	List optimization sensitivity values for each design parameter.
	History	Graph the history of the design goal, the mass, the stress, and each design parameter for optimization, or graph stress, displacement, mass monitors for parameter studies.
	Update Part/Model	Update the part to reflect the new geometry based on the optimization results.

Post Processing Task

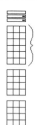

Post Processing task icons.

Icon	Command Equivalent	Description
Results	/RESULTS SELECT RESULTS	Select the results datasets to be displayed. By default, the first deflection and stresses are selected for display.
Display Template	/TEMPLATE	Select the type of display: contour, arrow plot, deformed geometry, etc.
Calculation Domain	/CALCULATION DOMAIN	Select the group of elements for the stress calculation domain. Set options for averaging across elements.
Display	/DISPLAY RESULTS EXECUTE	The *Display* command will execute the defined display. If the calculation domain is set to "selected elements," the program will prompt you to pick the elements to display. The default is to plot all the elements.
Probe	/DISPLAY RESULTS PROBE	Pick specific locations on the display to list the stress value. Note: in order for the *Probe* command to work, turn on the *Probe* switch in the *Display Template*.
Animate	/ANIMATE RESULTS	Animate the defined display.

Graph XY
 	
Setup XY Graph
	/GRAPH
	  XY GRAPH
	    SELECT ENTITIES
	    SET AXIS TYPE
	    STORE
	    GRAPH
		Create, store, and graph XY functions. For example, graph stress vs. distance along a line of nodes.

I-DEAS Visualizer	/VISUALIZER	Display results using a hardware-dependent display in a separate graphics window.

Beam Sections Task

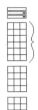

Beam Sections task icons.

Related icons are also included below that pertain to the display and processing of beams in other tasks.

Icon		Command Equivalent	Description
	Box Beam	/CREATE STANDARD SHAPE	Create standard shape beam sections.
	Pipe/Round Tube Beam		(Several other shapes are included, but not listed here.)
	Wide Flange Beam		
	Solid Rect Beam		
	General	/CREATE GENERAL SHAPE	Create General beam sections.
	Key In	/CREATE KEY IN	Create beam sections by keying in section property values.
	Get Section	/MANAGE SECTIONS GET	
	Store Section	/MANAGE SECTIONS STORE	

Meshing Task Icons (Beam Data and Display)

	Beam Data
	
	Modify Beam Data
		/ELEMENT
		    BEAM_GEOMETRIC_DATA
		      ORIENTATION
		      END_OFFSETS
		      RELEASES.
			Defines end releases and end offsets. Orients beam sections.

		/DISPLAY_OPTIONS
		    BEAM_SECTION
		      ON, OFF
			Turn the display of beam sections on/off.

Model Solution Task (Beam Output)

	Solution Set		Remember to request element forces in Output Selections!

Post Processing Task (Beam Options)

	Beam Post Processing	/BEAMS
		    CONTOUR ON X SECTION
		/BEAMS
		    FORCE & STRESS
		/BEAMS
		    CODE CHECKING
		/BEAMS
		    LINE ON X SECTION
			Display stresses on the cross section of a beam.
			Display Force/Moment, Stress diagrams on the beam model.
			Check beams using code checking.
			Display the shear stress next to the beam cross section.

Test Application

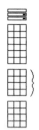

Test application icons.

Icon		Command Equivalent	Description
	Graph XY	/GRAPHS GRAPH XY GRAPH XY FUNCT	Graph a 2D Y vs. X function.
	Graph Stack	/GRAPHS SELECT GRAPH TYPE GRAPH XY	Graph multiple Y vs. X functions.
	Graph XYZ	/GRAPHS SELECT GRAPH TYPE GRAPH XY	Graph a 3D surface composed of multiple XY functions.
	Create	/FUNCTION OPERATIONS CREATE	Create functions.
	Single Math	/FUNCTION OPERATIONS SINGLE MATH	Math operations with one XY function argument.
	Multi Math	/FUNCTION OPERATIONS MULTI MATH	Math operations between two function arguments, such as add and subtract.
	Edit	/FUNCTION OPERATIONS EDIT	Edit functions.
	Individual Statistics	/FUNCTION OPERATIONS STATISTICS INDIVIDUAL	Perform statistical calculations.
	Correlation	/FUNCTION OPERATIONS STATISTICS CORRELATION COEF	Calculate correlation coefficient between two functions.
	Regression	/FUNCTION OPERATIONS STATISTICS REGRESSION ANAL	Polynomial regression of order 1-20 (1 = linear regression.)
	Manage Functions	/MANAGE FUNCTION	Manage function ADF records and arrays.

Modal Task

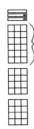

Modal task icons.

	SDOF Polynomial	/PARAMETERS
		  TECHNIQUE
		    SDOF POLYNOMIAL
			Select method of parameter extraction.

	Pick Frequencies	/PARAMETERS PICK FREQUENCIES	Pick frequencies off graph.
	Select References	/MANAGE MODAL REFERENCES	Select Reference DOF.
	Display References		Display Reference DOF on model.
	Display Responses		Display Response DOF on model.
	Calculate Residues	/PARAMETER CALCULATE RESIDUES	Calculate residue values for each parameter (frequency).

	SDOF Polynomial Shape	/SHAPE
		  TECHNIQUE
		    SDOF POLYNOMIAL
			Extract mode shapes for each function.

	Manage Parameter	/MANAGE PARAMETER (MODAL)	Open Parameter ADF file.
	Manage Shape	/MANAGE SHAPE (MODAL)	Open Shape ADF file.

Manufacturing Application

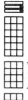

Manufacturing application icons.

The icons in this section of the panel are the same as the Design application.

Generative Machining Task

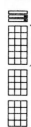

Generative Machining task icons.

Icon		Description
	Create Job	Create a job, giving it a name.
	Job Planning	Displays the Job Plan (Setup, OpGroup, Operations), and the Operation Sequence. Use this form to Add or modify Setups and (You can do most things starting with this form.)
	Add Part	Add parts to be machined to the job. (Prior to I-DEAS 8, a job was associated with one part.) Other icons in this stack add stock, clamps, fixtures, machine, etc. to the setup assembly.

Generative Machining Task (Cont.)

Icon		Description
	Add Setup	You can also find this icon on the Job Planning form to add a new Setup.
	Add OpGroup	Add an OpGroup to the Setup. You can also find this icon in the Job Planning Form.
	Add Operation	This command creates a new operation in the current OpGroup selected in the *Job Planning* form. (See below for details.)
	Process	Generate the toolpath for the operation.
	Animate	Animate the toolpath, showing the tool rotation.
	Show Next	Show next operation. (Similar icon to show previous.)
	Write CL File	Output the Cutter Location File.

Operation Specification Form Icons

After you select the type of operation, the Operation Specification form defines the details of the Operation. This form contains the following icons in a row from left to right:

Icon		Description
	Pick Geometry	Pick the geometry to be machined.
	Coordinate Systems	Define operation machining coordinate system, if different from the machine coordinate system.
	Boundaries	Pick sections to machine within. Used for Mill operations.
	Tool Specific- ation	Define tool type and geometry.
	Machining Parameters	Define cutting parameters.
	Process	Generate the toolpath.

Assemble Setup Task

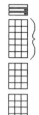

Assemble Setup task icons.

This task is similar to the Assembly task in the Design application, with the addition of specific machining instances in the assembly.

Icon		Menu Equivalent	Description
	Hierarchy	/ASSEMBLE HIERARCHY	Define assembly hierarchy. Add parent to start a new Assembly.
	Add Stock to Setup	/ASSEMBLE ADD...	Add instances of stock part to the setup assembly. There other similar icons to add machine, fixtures, and clamps to the setup.

Appendix B
Advanced Features and Interfaces

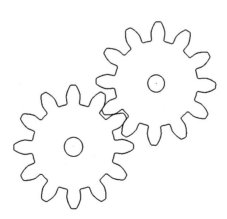

Introduction

This chapter covers advanced features of the main applications of I-DEAS such as Design, Simulation, Manufacturing, and Test. These topics are not required for beginning users of I-DEAS, but can help more advanced users be more effective with the software. Some of the features described below will be slightly different in the Drafting application, since this application has some differences in its user interface.

Display Capability

So far, this guide has only used some of the basic display capability using line, hidden line, and shaded options, using one viewport. More advanced options are available, with the screen divided up into multiple areas, called viewports.

Advanced Imaging

Most terminals and workstations are capable of displaying shaded image displays. In Software Shading mode, the software computes the raster image, based on the Display Options of lighting and other switches, and the Display Attributes of Glossiness, Brightness, Radiant light, Diffuse light, and Translucency. This basic shading capability does not include shadows or reflections.

In Hardware Shading mode, the shading is computed in the hardware, on terminals that support this feature. These displays also do not support shadows and reflections, but are much faster than doing the same thing in software.

In the Advanced Shading mode, the shading is computed in software, but using a more advanced "ray tracing" algorithm. This algorithm can include shadows and reflections. These can be turned on or off with switches in the software. The number of reflections to consider is an option. Some users feel that two to four reflections are required to make very realistic displays of shiny objects. Other options available in advanced shading are Anti-aliasing, which makes edges appear less jagged, and Pixel Filtering, which makes the display more coarse by grouping pixels together.

Keep in mind that using advanced shading can be time consuming. The time to generate these images will also be greatly affected by the number of the "fancy" options that are turned on. You will probably need to make more than one image to get all the options the way you want. Some things that can help to minimize the time are to use a smaller viewport, and to use a large value for pixel filtering until you get the options and lighting set satisfactorily.

Texture Mapping

Some computer hardware has the capability of adding realistic texture mapping to the surfaces of parts. The option to turn this on is found on the Appearance form when a surface or a part is selected. Various textures can be selected on the form. These options will not work with X3D display or with all hardware graphics.

Multiple Viewports

The graphics window can be subdivided into up to four viewports. These four viewports can be used at the same time, such as to display a model with different viewing angles or display options in each. The definition of the layout of these viewports is manipulated with a stack of icons in the lower section of the icon panel.

 One Viewport

When using multiple viewports, dynamic viewing can be used in any of the displayed viewports by placing the cursor in the viewport you want to change. Some commands may ask an additional question to request which viewport or viewports to take the requested action in, or will only act on the "active viewport." One of the icons in this stack will let you define one viewport as the "active viewport."

Another reason for redefining viewports is to define one viewport smaller than normal. This can be useful on some computers to keep the graphics away from areas on your screen. Using a smaller viewport will also speed up advanced shaded image displays. This is useful as you experiment with options.

Annotation Overlays

Another capability in I-DEAS is the ability to add your own annotation to any graphic display. This includes text, lines, arrows, boxes, and circles. For example, in many of the workshops in this Student Guide, the extra information added to the I-DEAS graphics to indicate which points you were supposed to pick were generated using an annotation overlay.

All the annotation added to the screen can be thought of as a transparent sheet with your annotation on it. These annotation overlay "sheets" can be managed (stored) like other entities in I-DEAS.

 Screen Annotation

Picture Files and Printing

There are two ways to produce hardcopy of your screen images. The first method is to use screen capture utilities. These sometimes come as part of the operating system of the computer. The second method is to use utilities in the software to store picture files and to print them.

Picture Files

Picture files are files containing I-DEAS images. They are used for saving the image for later display or plotting. Anything displayed in the graphic area can be stored in a picture file for later display. Displaying from a picture file will be much faster than calculating the image the first time, since the computation required to generate the picture has already been done.

To create a picture file, turn the picture file write switch on, either for single pictures or for multiple pictures. When an image is created with either switch on, the program stores the picture in a file. With the "Create Single" switch on, the picture will be stored in the filename supplied. With the "Create Multiple" switch on, the filename will have a new number appended to it for each new picture stored. If you want to make a picture file of an image already on the display, use the Redisplay command to redraw the image and simultaneously write it to the file. The write switches are turned on with the commands:

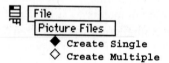

There are several different format options for the file that will be written. The most general type is "formatted." This is an ASCII file that can be edited or transferred between computers. The next type is "binary." It is terminal independent, but may not be computer transportable, since different computers may use different binary file structures. The last type is "replay." This type is terminal specific. It is the specific escape sequences and instructions for the particular terminal you are using. The default file type is binary.

Picture files are displayed with the command:

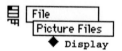

Printing

Printing is done through the picture files described above. Picture files can be converted and sent to a plotter from within I-DEAS using the command:

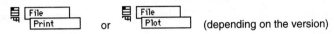   (depending on the version)

 In I-DEAS 8 or higher, there is a one-button print capability with the icon *Quick Print*. To use this command:

1. Your I-DEAS Administrator must first run the *printAdmin* utility to setup the printers.

2. You then run *File*, *Print* to select the default I-DEAS printer.

3. You may then use the *Quick Print* icon to directly print from the workbench, or use *File*, *Print* to print either from a picture file or from the workbench.

Screen Captures

On Windows, the Print Screen key on your keyboard can be used to capture entire screen, or use Alt-Print Screen to capture just one window such as the Graphics window to the clipboard. You may then paste this image into other applications.

Unix workstations may have their own screen plotting independent of I-DEAS. On some systems, the keyboard command HC (Hard Copy) in I-DEAS or the *Print Screen* Key will be programmed to trigger a local screen capture routine supported by the local hardware. Some plotters hook up directly to the RGB (Red, Green, Blue) terminals in the back of the workstation. I-DEAS does not know anything about this kind of plot interface, so this will be described in your computer system manuals, not in any I-DEAS documentation.

Standard UNIX utilities to capture images within a window are xwd or xv. Again, this is entirely independent of the I-DEAS software.

Exporting IGES Files

IGES is a standard way of exporting part geometry to interface with other programs that need part surfaces or boundary edges. A 3D part can be exported from the Modeler task using the menu *File*, *Export* to export a view-independent IGES file. A 2D drawing can be exported from the Drafting task.

A typical use for a 2D IGES file would be to interface with a numerical controlled (NC) milling machine or another type of 2D manufacturing. You may export a simple 2D profile of a part using the Drafting task. Create a drawing of the part with no borders. Hide or delete dimensions so that they won't be exported. Then export the IGES file.

To export a 3D part, use the *File*, *Export* menu in the Design application. Either pick individual surfaces of a part, or the entire part. (It is assumed that you can recognize the graphic feedback to determine which is selected.) In this way, you can export a file containing every surface of the part or just selected surfaces.

When you read a 3D IGES file from another CAD system, there are sometimes problems with the quality of the data due to the tolerances used by different CAD programs. Display free edges to find surface edges that are not stitched together. Use the other checks in the *Diagnose Part* command to check for invalid geometry or duplicate surfaces.

Depending on your use for the imported geometry, you may need to take different steps to clean up or repair the bad geometry. If your use for the imported geometry is to create a finite element model, you may use the section meshing capability to create a mesh of elements over unstitched surfaces. If your use for the imported geometry is machining, the newest version of the Machining application may tolerate the unstitched surfaces. In other cases, you may have to delete bad surfaces and rebuild them using tools like *Surface By Boundary*. There are some tips in the online help manuals on methods of cleaning up imported IGES geometry.

Programmability

Programmability is a powerful feature to allow you to customize the program for your own special applications. The word "programmability" refers to several different related features in the I-DEAS software. These include the ability to define variables, create program files with regular I-DEAS commands plus special programmability commands, the ability to define your own symbols to be equivalent to a string of commands, or to program an icon to run a sequence of commands.

Variables

At any prompt for information in the software, you can type a FORTRAN-type expression rather than just entering a number. This expression can contain variables. To define variables, type a pound sign (#) at the start of the line as shown:

```
#WIDTH=10.3
#LENGTH=WIDTH*5
#HEIGHT=WIDTH+.25
#VOLUME=WIDTH*LENGTH*HEIGHT
```

Variables can contain numbers or character string data. To indicate the difference, strings must be surrounded with quotes. Strings can be concatenated together like adding numbers. Some sample statements using strings are shown:

```
#FILENAME="INPUT"
#FILENAME=FILENAME+".UNV"
#PARTNAME="X-"
#PARTNAME="X"+N
```

Variables can be singly dimensioned (one subscript). Variable names can contain up to ten characters. Some valid variable definition statements are as shown:

```
#DECLARE VALUES(25)
#VALUE(1)=3.0
#VALUE(2)=VALUE(1)*.5
```

So far, we have been talking about user defined variables. There are also some special predefined variables, depending on which module you are in. All predefined variable names begin with the characters "Z_". For example, the variable Z_PI is defined as the value of Pi. These variables are called program defined variables.

A third type of variable is a rolling stack containing the last 100 numbers listed to the screen in reverse order. This provides an easy way to get information into variables in your own programs. The name of this variable is Z_LIST. For example, if the last three numbers listed to the screen were coordinates that you want to capture into your own variables X, Y, and Z, you might use the following:

```
#X=Z_LIST(3)
#Y=Z_LIST(2)
#Z=Z_LIST(1)
```

Variables can be listed using the keyboard command LV (List Variable). All three types of variables above can be listed.

```
LV-LIST_VARIABLES
    USER
    PROGRAM
    STACK
```

Variables you have defined will be stored in your model file if you Save it. To delete them you can use the #DELETE command.

```
#DELETE FILENAME
#DELETE HEIGHT,LENGTH,WIDTH,VOLUME
#DELETE ALL
```

Special variables are also defined by some commands such as the *Measure* command. The variables listed in the List Window, such as DX1=123.456789, can be used to enter values in subsequent commands. To enter this value, you can enter the variable name DX1 when prompted for any numerical value. Or, for example, you may want to enter half this value as DX1/2.

Programmability Commands

All of the statements used to define variables above were started with a #
sign. This is true for all I-DEAS programmability commands. The # tells
the software that this is a programmability command as opposed to a
regular interactive menu command.

To write general programs, there are programmability commands such as
input, output, and program control (looping and branching).

The form of the #INPUT command includes a prompt string, the variable
that the response is read into, and an optional default response. The form
of the prompt that the user sees showing the default answer will look just
like any other prompt in I-DEAS. A sample of some variations on the
INPUT command are shown.

```
#INPUT "ENTER THE FILENAME" FILENAME
#INPUT "ENTER THE WIDTH" WIDTH 3
#INPUT "ENTER THE HEIGHT" HEIGHT HEIGHT
```

The OUTPUT command may be used to output text and values to the
screen as shown:

```
#OUTPUT X,Y,Z
#OUTPUT "THE TOTAL VOLUME IS ", VOLUME
```

Program control statements consist of statement labels and commands to
branch to statement labels. For example:

```
#BEGIN:                              (Statement Label)
#IF(A EQ B)THEN GOTO BEGIN           (Conditional Jump)
#IF (COUNT LE 0)THEN #EXIT           (Exit program or
subprogram)
#IF (FILENAME NE "NONE")THEN /MA WRITE UNIV
FILENAME
#GOTO BEGIN                          (Unconditional jump)
```

When a program file runs interactively, it will look just the same as if the
user were typing in the commands. A way to prevent the menus and
program inputs from being displayed is with the ECHO control statement:

```
#ECHO NONE    (Don't show commands from program)
#ECHO LIST    (Show only list displays)
#ECHO ALL     (Show all commands, lists, and menus)
```

Program Files

A program file can be created interactively, with an editor, or with a combination of the two. To create a program file interactively, issue the command:

```
PROGRAM FILES
     CREATE
          Filename  XYZ
```

Everything you type, or cursor locations you pick, will be entered into this file. To make the file more readable later, it is suggested that you type commands more completely than required, for example to type "CREATE" rather than just "C". (An exception to this if you want to create programs that will run with different local language interfaces to the software. Since the mnemonics for commands are the same in each language, you should use these mnemonics to create "international" programs.)

When you want to end writing commands to the file, give the command:

```
PROGRAM_FILES
     END_CREATE
```

Programs can be executed using the command:

```
PROGRAM_FILES
     RUN
```

Simple, straight-through program files can easily be created interactively. Even relatively sophisticated programs which include program control statements can be created interactively, since statement labels and branching statements can be typed interactively, even though these statements have no effect then.

Quite often, the most efficient way to create a program file is to initially create it interactively, and then use an editor to add looping, branching, and comments, or to fix errors. When you list or edit a program file, you will notice that every line begins with a prefix that you did not see inside of the software. A "K :" shows that this line came from the keyboard; an "E :" marks the end of file. You can add your own comments using a "C :". These prefixes are required in the first three spaces on a line.

Another use for program files is to get any information in the List Window into a file. When you create a program file, there are several options to control the information that goes into the file. If you turn on the List switch, everything going to the List Window will also be recorded in the Program File.

The other options on the form are the Prompt switch to capture the text of the program prompts and the Menu switch to capture the menu choices in the program file.

Global Symbols

Global symbols are a simple form of programmability where the user can define a symbol name that can be used to issue a string of commands. Don't confuse "global symbol" with "global command." Both are global meaning that they work at any level of the current menu. The word "symbol" means that this user defined name symbolically stands for something else. Another word for this is *alias*.

For example, a global symbol named SHOW that will display picture files could be defined as follows:

```
/OPTIONS
    GLOBAL_SYMBOLS
        ENTER
            Symbol name?# SHOW
            Contents# "/FILE PICTURE"
```

The contents of a global symbol should begin with "/" so that the commands work properly no matter where you were in the menus when the symbol was given. It is also a good practice to enter the entire contents surrounded with quotes as shown above to make sure that any special characters or variable names in the symbol are not interpreted right away.

User Defined Icons

User Defined Icons are programmed using the same techniques as defining global symbols. The only difference is that the user picks the icon with the mouse instead of typing a key word at the keyboard.

The icon *User Panel* will display a sub-panel that can be user defined. This sub-panel will contain the icons U1 through U6. To program these icons, just define a global symbol with the names "USER1" etc. For example, if you define a global symbol named USER1 instead of SHOW as described above, the icon U1 will then bring up the picture file form.

The global symbols USER1, etc., could be programmed with a simple string of commands, sometimes called a macro, or they could be programmed to run a more extensive program file.

User Interface Features

Some additional user interface items are included here. For more complete information, read the Help Library information indicated.

Keyboard Accelerators

Some commonly used commands have control key short-cuts. When you pull down a stack of icons or commands from the menu bar, these keystrokes are shown on the right. For example, ^S and ^O can be used to Save and Open model files. ^Z reopens the current model file as of the last Save. ^L can be used to execute a line mode display. For a complete list of these keyboard accelerators, select

Help, Help Library
I-DEAS User's Guide
Special Functions
Global Commands and Keyboard Accelerators

Report Writer

Report Writer is a general tool that lets you create reports using your own customized formats. Several different parts of the software use this utility to allow you to generate custom reports. It is available in the data management forms: Manage Items, Manage Libraries, and Manage Projects. It is also available to print results in the Simulation application.

As an example, a project manager could use this utility to create reports on the status of parts created by team members.

Relational Data Manager

I-DEAS includes a relational database management system the Management application, the Relational Data Manager task. This system can be used to create, manipulate, plot, read, write, and store tables of information.

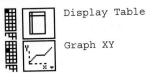

Display Table

Graph XY

This capability in I-DEAS has many uses. Many of the interfaces to other external software have been written using this capability. It also can be a tool to directly manage and plot tabular data in any way you like. This capability, along with I-DEAS programmability and variables, allows sophisticated user application programs to be written in the I-DEAS environment.

Start the Relational Data Manager task with the menu-bar command:

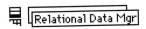

The steps to use the Data Manager system are:

1. Open a database.
2. Create a table "schema."
3. Enter data into the table.
4. Manipulate and plot the table.
5. Store the table.
6. Close the database.

The database opened by this task is not the same as the model file. It is a separate file. It is "saved" by closing the file. Multiple tables of information can be stored in the same file. When a table is created, you must first define the "schema," which is the column names and the type of data in each column. To permanently save a table, you must *Store* it in the database and also *Close* the database.

The Relational Data Manager contains commands to modify tables by adding information or by performing operations on single or multiple tables such as sorting, selecting, or comparing. It can also be used to read and write to ASCII files with its Translate commands. This can be used to write interfaces to get data into or out of I-DEAS.

More information on this capability can be found in the online documentation. To find a discussion of this topic, see the online manual "Relational Data Manager User's Guide."

Summary

This chapter has introduced some slightly more advanced capabilities of the user interface. A very unstructured workshop follows, suggesting some activities for you to use to investigate programmability and variables. Some of the previous workshops showed you how to record your commands in a program file. These program files can be more powerful when you learn to add prompts and variables to them to perform the same operations with different parameters.

·Ö· PARTING_SUGGESTIONS

-When experimenting with advanced imaging, use a small screen size to test options.

-Use *Help*, *On Context* to investigate other options available.

Where To Go For More Information

The material presented in this chapter is covered in the online documentation. Look in the following locations:

Help, Help Library
  I-DEAS User's Guide
    Global Functions in I-DEAS
      Getting Started With I-DEAS
        Setting the Display Style
      General Capabilities
        Surface Appearance
        Light Sources
        Clipping Methods
        Advanced Shading
        Color Use in I-DEAS
      Special Functions
        Importing and Exporting Files
        Launching Other Applications From I-DEAS
        Creating Reports (With Report Writer)
        Adding Annotation Graphics
        Troubleshooting I-DEAS
        Global Commands and Keyboard Accelerators
        Recording Macros (Program Files)
      Program Files
        Working With Program Files
      Picture Files and Printing
        Using Picture Files and CGM Files
    Geometry Data Translators User's Guide
    File Formats Reference Guide
    Relational Data Manager User's Guide
    Open I-DEAS User's Guide
    General I-DEAS Administration
      I-DEAS Print Setup and Administration
    I-DEAS Print User's Guide
    I-DEAS Print Administration Guide

Workshop: Advanced Features

This workshop will give you a chance to experiment with some of the advanced features of the I-DEAS software such as programmability, variables, and the Relational Data Manager task.

Before you start:

1. You should be familiar with the basic functions of the I-DEAS software.

After you're done, you should be able to:

1. Create and display picture files.
2. Create global symbols and variables.
3. Extract values from the "stack" using the Z_LIST variable.
4. Create and run program files.
5. Create and graph a table in the Relational Data Manager.

Workshop Instructions:

Enter I-DEAS with a new model file. Enter the Design application, Modeler task.

 ...

Create a part and display it on the screen.

 ...

Picture Files

Store this picture in a picture file.

 File
Picture Files

◆ **Create Single** Filename: **Picture1**

 ... *(Execute some type of display like Redisplay or Line.)*

Put away the part, and then display the picture file you created.

 File
Picture Files

◆ **Display** Filename: **Picture1**

Global Symbols and User-Defined Icons

Create a global symbol named "USER1" to display this specific picture file. Turn on menus to create the global symbol.

Menus ■ Menu Display On ◆ All

```
/Options
    Global_Symbol
        Enter
    Enter Global Symbol name USER1
    Enter Global Symbol definition    Type this on one line:
/file picture_file picture_file_mode display filename picture1 ok
```

Either typing "USER1" or pressing icon U1 in the user panel will execute this command string to display this picture.

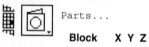 User Panel

Create a global symbol named "USER2" to output some text to the list window.

```
/Options
    Global_Symbol
        Enter
    Enter Global Symbol name USER2
    Enter Global Symbol Definition #output "Hello There"
```

Press user panel icon U2 to execute this command.

User-Defined Variables and the "Stack" variable.

Define variables, named X, Y, and Z, and list them using the List Variable keyboard command. (Type variables in the prompt window as shown below, including the "#" as shown.)

```
#X=5
#Y=6
#Z=3
LV (List Variables. Type this on the keyboard.)
    USER
```

Create a block from the parts catalog using these variables X,Y, and Z as the dimensions of the block.

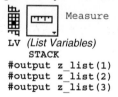 Parts...

Block   X Y Z

Measure the distance between two points, and then list the Z_LIST stack to see how these numbers are stored. List variables using the #OUTPUT statement. The use of the stack variable allows your programs to capture the output of any commands that list information to the List Window.

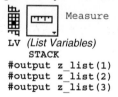 Measure

```
LV (List Variables)
    STACK
#output z_list(1)
#output z_list(2)
#output z_list(3)
```

Program Files

Create a program file which will prompt you for the X,Y,Z dimensions and then create another block using these variables. (Turn on menus.)

    ☐ | File
        Program_File
            Create
                XYZBLOCK

#X=2
#Y=3
#Z=4
#input "Enter X Dimension" **X   X**
                (Take the default answer now.)
#input "Enter Y Dimension" **Y   Y**
#input "Enter Z Dimension" **Z   Z**

Options, Preferences, Forms
☐ Forms Display *(Off)*
|OK| |OK|

    ☐ | 🗂 | Parts...
Catalog Item
Enter index of item # (Done) **1**   *(Block)*
Enter index of item # (Done) **<Return>**
Parameter
Enter index of item # (Done) **1**   *(Length)*
Enter index of item # (Done) **<Return>**
Value
Enter new value (100) **X**
    (Repeat using Y and Z variables for Height and Depth dimensions.)
Okay

Options, Preferences, Preferences Menu
Enter index of item # (Done) **2**   *(Forms)*
Enter index of item # (Done) **<Return>**
Forms Display Switch
Enter forms display enable switch ON or OFF(on) **<RETURN>**
Okay
Okay
    ☐ | File
        Program_File
            End_Create

Run the program file.

    ☐ | File
        Program_File
            Run

Program user panel icon U3 to run this program file.

    /Options
        Global_Symbol
            Enter
        Enter Global Symbol name **USER3**
        Enter Global Symbol definition **/file pr run fil XYZBLOCK ok**

Press user panel icon U3 to execute this command.

     User Panel

Sample Program File - Involute Gear

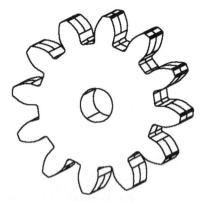

The pages to follow present a sample program file to create the wireframe geometry for an involute gear.

The geometry of the involute gear tooth form is defined by the number of teeth, the pitch radius, and the pressure angle. Other parameters are the base radius used to generate the involute, and the inner and outer radii. From these parameters, the program will create the wireframe geometry of one half of a tooth.

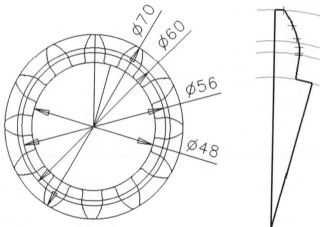

This geometry can be used to create a complete gear, as shown.

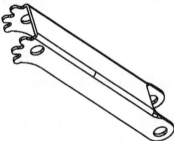

Although you could create a program to create a complete gear, it may have more utility to create just the wireframe geometry. For example, the part above is from a car jack mechanism, used for the linear buckling tutorial in Chapter 9. This mechanism uses a segment of an involute gear, but not the complete gear. The program files presented here could be used to create the wireframe geometry for this feature.

Section 1 - Input variables and create circles and lines.

This segment of the program illustrates how to use variables and ask the user for input. It also demonstrates how to create some geometry based on the variables in the program.

The comments in the program are optional.

The program only prompts the user for the required input and computes the other parameters. Notice also that the user is given a default answer in some cases, such as for the pressure angle and the inner radius.

```
C : Demonstration program file to create
C : involute gear geometry
C : I-DEAS Student Guide
C :
C : RP=Pitch radius
C : RB=Base radius of cirlce
C :    used to generate involute
C : RO=outer radius
C : RI=inner radius
C :
K : #PI=3.14159265359
K : #input "Enter number of teeth:" N
K : #input "Enter pitch radius:" RP
K : #input "Enter pressure angle:",AP 20
K : #AT=360/N
K : #P=N/RP/2
K : #RI=RP-1.157/P
K : #input "Enter inner radius:" RI RI
K : #RO=1/P+RP
K : #RB=RP*COS(AP)
C :
C : Create circles
C :
K : /cr c ce
K : OP
K :   CX 0 CY 0 RD RO
K : APPL
K :   RD RI
K : APPL
K :   RD RP
K : APPL
K :   RD RB
K : OKAY
K :
C : Create radial lines
C :
K : /cr l si
K : OP
K :   FX 0 FY 0 LH RO LA 90
K : APPL
K :   FX 0 FY 0 LH RI LA 90-AT/2
K : OKAY
K :
```

To see how this program file works, you could either type it in as a text file and run it, or type the commands interactively to create it.

It will be easier if you get this segment working first, then put it together with the segments to follow.

Section 2 - Loop to create points on the involute profile

This section of the program does the "hard" part to create a number of points on the tooth profile. A loop is used to create multiple points. The variable "NP" could be changed in the program to change the number of points created.

```
C  :  Create spline points
C  :
K  :  #A1=90-AT/4+AP-RP*SIN(AP)/RB*180/PI
K  :  #A2=SQRT(RO*RO-RB*RB)/RB*180/PI+A1
K  :  #NP=6
K  :  #AD=(A2-A1)/(NP-1)
K  :  #I=1
K  :  #A=A1
K  :  #SPLINE:
K  :  #L=RB*(A-A1)*PI/180
K  :  #X=RB*COS(A)+L*SIN(A)
K  :  #Y=RB*SIN(A)-L*COS(A)
K  :  #OUTPUT I," X=",X," Y=",Y
K  :  /cr p
K  :  OP
K  :    FX X FY Y
K  :  OKAY
K  :
K  :  #A=A+AD
K  :  #I=I+1
K  :  #IF(I LE NP)THEN GOTO SPLINE
C  :
C  :  Create line at base of tooth
C  :
K  :  /cr l si
K  :  OP
K  :    FX RI*COS(A1)  FY RI*SIN(A1)
K  :    SX RB*COS(A1)  SY RB*SIN(A1)
K  :  OKAY
K  :
```

To complete a spur gear, create a spline connecting the points, and extrude the section. Fillet the root and the tip. Reflect to create a complete tooth. Use a pattern to complete the gear. Cut a hole for the center shaft.

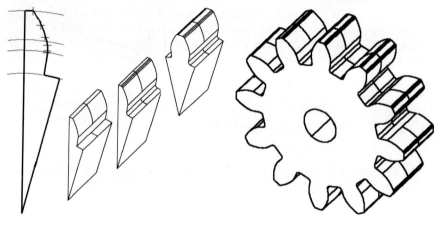

Section 3 - Create a spline using an equation

This section of the program is optional, because the previous programming segments generated enough geometry to create a gear. This program segment shows how to create the involute spline using an equation instead of manually fitting a spline between points.

It also demonstrates a clever technique of a program file creating another program file, and then running it.

```
C : Create function spline equation
C :
K : #u1=a1*PI/180
K : #u2=a2*PI/180
K : #OPEN FILE1 "fs.prg"
K : #write FILE1 "K : /cr m3 f"
K : #write FILE1 "K : AE"
K : #write FILE1 "K : NAM involute tooth"
K : #write FILE1 "K : "
K : #write FILE1 "K : FRO ",u1," TO ",u2
K : #write FILE1 "K : EQN APPEND"
K : #write FILE1 "K : U1=", U1
K : #write FILE1 "K : RB=", RB
K : #write FILE1 "K : L=RB*(u-u1)"
K : #write FILE1 "K : X=RB*COS(u)+L*SIN(u)"
K : #write FILE1 "K : Y=RB*SIN(u)-L*COS(u)"
K : #write FILE1 "K : Z=0"
K : #write FILE1 "K : "
K : #write FILE1 "K : DONE; OKAY; OKAY; CANC"
K : #write FILE1 "E :"
K : #close FILE1
K : /f pr r FIL fs.prg OKAY
K :
K : LV U
E : **** END OF SESSION ****
```

A limitation of this program segment is that it won't work twice as written because it does not delete existing equations first. To run it a second time in the same model file, first delete the function spline named "involute tooth" created by the program.

After you run the program file above, you will see the program "fs.prg" in your directory that it created, with contents that looks like the following. It does not delete this file when it ends, but it will overwrite it if you run the program file a second time.

```
K : /cr m3 f
K : AE
K : NAM involute tooth
K :
K : FRO 1.424992 TO 2.160807
K : EQN APPEND
K : U1=1.424992
K : RB=28.19078
K : L=RB*(u-u1)
K : X=RB*COS(u)+L*SIN(u)
K : Y=RB*SIN(u)-L*COS(u)
K : Z=0
K :
K : DONE; OKAY; OKAY; CANC
E :
```

All programs should terminate with the "E :" in the last line.

Relational Data Manager

To experiment with the Relational Data Manager task, switch to the Management application and the Relational Data Manager task.

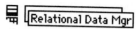 Relational Data Mgr

Define a table with the columns "Part ID", "Length", and "Weight".

```
/Database
    Open
        Name: JUNK
/Create
    Enter name of column 1 Part ID
    Enter data type Character
    Enter number of characters in each column 20

    Enter name of column 2 Length
    Enter data type Real
    Enter number of values in each row (1) [OKO]
    Enter Class of units for column 2 (None) [OKO]

    Enter name of column 3 Weight
    Enter data type Real
    Enter number of values in each row (1) [OKO]
    Enter Class of units for column 3 (None) [OKO]

    Enter name of column 4 [OKO]

    *** Now entering data in Row 1
    Enter Part ID  Part Number 1
    Continue (End_Character) [OKO]
    . . .  (Enter some fictitious numbers for several rows.)
```

Display this table, and graph column 3 vs column 2.

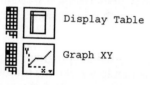

Display Table

Graph XY

Appendix C
Troubleshooting Guide

Troubleshooting Methodology

When errors occur, it is important to follow a logical troubleshooting methodology to identify and solve the problem. What was the first symptom of the problem? Under what consistent conditions does the problem occur? Under what conditions does it not occur? Can you narrow down the problem to the smallest set of conditions to reproduce the problem? Knowing these conditions, is there another way to attack the task to avoid the problem? With the problem isolated as much as possible, can you work around the problem yourself, or do you need Hotline support? Should the problem be reported so that it can be fixed in a future release?

Problems can be caused by various root causes, including:

- •Physical hardware problems
  RAM memory errors, bad circuit boards, loose connections,
  worn out magnetic disk media,
  files corrupted by hardware errors, etc.

- •Hardware installation problems
  not enough RAM or disk memory,
  disk swap space not partitioned properly,
  network parameters not properly configured,
  a different operating system version than specified,
  operating system patches need to be installed, etc.

- •Software installation problems
  Software security not running or not accessible,
  team directories/projects not configured properly,
  not all files loaded or accessible from your workstation,

- •Software errors
  User has requested conditions not anticipated or tested,
  not enough memory is available to the software,
  other hardware resources not sufficient, etc.

- •User input errors
  Requesting values outside of reasonable ranges,
  incompatible combinations of inputs,
  modeling errors too small to be seen, or non-physical, etc.

The more that you can test various conditions above to isolate the root cause of the problem, the quicker you will solve it and get back to your work.

This troubleshooting guide should help you learn how to identify problem causes, where to call for support help, and how to report problems so they can be fixed.

Investigating the problem

First, do some detective work to investigate the problem. Computer software is usually very deterministic, meaning that the same input variables always yield the same results. You may not be aware of all of the input variables, however. Consider other variables that might be having an effect, such as operating system parameters, kernel parameters, network load, memory in use by other programs or users logged on to your workstation, etc.

Collect available clues. These include error messages, error output files, messages written to the list window, any other messages, or other beeps or extraneous noises the computer is making.

Try different conditions one at a time. Try to minimize the differences in each test. Stop other programs running on your workstation, and turn off unnecessary desktop accessories like clocks and other utilities that may be using memory or "stealing" colors from the software. Do other users have the same problem? Do other users running from a different account on your workstation have the same problem? Try to identify every possibility and rule them out one at a time. If another user on the same machine does not have the problem, it must be something different about each user's environment, such as the variables defined in a startup script. On the other hand, if you can reproduce the problem on another workstation, the problem must be due to something they both have in common.

Possible problem conditions that should be investigated and ruled out one at a time include these categories:

- Network hardware and parameters (the way disks are mounted)
- Operating system version and parameters
- Workstation hardware - memory and disk, graphics boards
- System and user-defined variables (.profile, desktop utilities, etc.)
- I-DEAS Data Management (data installation, project)
- I-DEAS model file
- I-DEAS software (specific version)

Intermittent problems are always the most difficult to diagnose. The problem may only seem intermittent because you are not aware of all the variables in the equation. Intermittent problems may not affect other software if it uses less memory or writes to different RAM memory or disk locations.

Later sections in this troubleshooting guide will identify possible causes for a variety of typical user problems. After investigating these basic problems, the next step would be to call the Hotline. Even if you can't find a solution, you will have more information to document your problem.

How to get technical support

The GTAC (Global Technical Access Center) is the focal point for technical questions. http://support.plms-eds.com/ You will need to know your customer ID number.

Before you contact GTAC, follow through the steps above to identify the significant related conditions. When you call the Hotline, have this information ready:

- Customer ID number.
- What hardware you are using - manufacturer, model?
- What version of the operating system you have?
- What version of the software you have?
- What application and task are you using?
- Brief description of the problem.

Be able to describe the simplest steps that can be taken to demonstrate the problem. Be ready to answer questions like:

- Does the problem occur on other workstations?
- Does the problem occur for other users on the same workstation?
- Does the problem occur in a new model file?
- Does the problem occur on other types of workstations?
- Does the problem occur when using other graphics drivers?
- Is this a new problem where it used to work before?
- Has anything else changed recently? (System upgrades)

These are all important clues in diagnosing the problem.

If you can report the problem using e-mail rather than by telephone, it will be more efficient, because you will be forced to concisely describe the problem conditions, and it will give the support engineer time to research the problem.

The end result of a call to the Hotline should be one or more of the following:

(1) a solution to your problem,
(2) a work-around to complete your task another way, or
(3) a bug in the software that needs to be reported.

How to submit a Problem Information Report (PIR)

Whether you solve your problem yourself, or with the help of the Hotline, if the problem is a result of a software bug, it should be reported so that it can be fixed. The reporting mechanism to report these errors or to submit enhancement requests is called a Program Information Report (PIR). It can be submitted through the Hotline or through a local Application Engineer (AE).

To submit a PIR, all of the information about the hardware, operating system, and I-DEAS version are necessary.

Running the software

The first category of problems involve getting the software running at all or reading or writing your model file.

These problems are generally due to hardware configuration or software installation. This category does not include general user interface or specific software application problems which will be discussed later.

Software won't start

Why won't the software start?

Diagnostic questions:

Does any I-DEAS startup message get sent to the window or does it just respond "command ideas not found"?

How far does it get? Does the *Start Form* come up or does it stop before that?

Is a file created named error1234.out? (1234 can be any number.) If so, what is in this file? (List the file.)

What happens if you run I-DEAS with the X3D driver?
(Type: ideas -d x3d)

Possible causes:

Your "path" is not defined to include the ideas "bin" directory.

You don't have enough RAM or swap space to run I-DEAS.

You don't have enough disk space to create a model file.

You don't have file permissions to create a model file. (Are you working in your home directory?)

You are trying to start I-DEAS with a specific hardware graphics driver, but you are running remotely or are using an X terminal. Try running I-DEAS using X3D.

The software security was not properly installed or has timed out.

For more information, see:

General I-DEAS Administration Guide for Unix (Online)
General I-DEAS Administration Guide for Windows NT (Online)
I-DEAS Installation Guide (Print only, not online)
I-DEAS Site Planning Guide (Print only, not online)

Can't open a project

After you type "ideas," or select the menu from the Start menu, the I-DEAS Start form is displayed. The first thing you enter on this form is the project name. You either enter a new name to create a new project, or select an existing name to open an existing project.

Why can't I create a new project?

There are two reasons you might not be able to create a new project. The first reason is that the system administrator who installed the software might have set it up so that individual users cannot create projects. This is typical in large companies, where the project structure is defined ahead of time. The second reason is that the main file that keeps track of the list of projects has somehow become corrupt, or has the wrong file permissions so that a new project cannot be added to it.

Why can't I open an existing project?

There are also two main reasons why you might not be able to select an existing project. First, the project may have been configured to only allow certain project members to access it. Second, the project file that keeps track of all the information in the project may be either corrupt or have the wrong file permissions for you to access it.

For more information, see:

Help, Help Library
  Data Management User's Guide
    Troubleshooting I-DEAS Data Management (IDM)
      Troubleshooting project problems
      Troubleshooting library problems
      Troubleshooting model file problems
      Troubleshooting file access problems
      Troubleshooting miscellaneous problems

Can't read/write to your model file

When you open a model file, the software creates scratch files that collect the changes you make to the model file. These scratch files contain a copy of any block of data that differs from the original model file. When you give save, the changes are written to the model file that you opened.

Why can't I create a new model file?

If you cannot create a new model file, it could be a problem with writing the scratch files, such as a problem with file permissions, or it could be that you do not have enough file space on your disk to create the scratch files.

The location where you are trying to write the scratch files is usually your default directory. If you have changed your default directory by using the "cd" command, you may have set your default directory to a location where you do not have permission to write files.

Why can't I open an existing model file?

If you cannot open an existing model file, it could either be a problem with reading the model file or writing the scratch files. It could be a problem with file permissions in either case or insufficient file space on your disk to create the scratch files, as mentioned above.

If the software cannot read the contents of the model file, usually little can be done. The best strategy is to understand what causes this to occur and adopt practices the protect yourself against this kind of data loss.

What can cause this problem? The model file can become corrupted if a crash or interrupt occurs during the save operation. This is unlikely, but happens since things are mechanically moving when the disk is writing. Also, magnetic media can have errors. The disk drive containing the file could have been mounted with the wrong parameters. In most cases, all disks that the I-DEAS software uses should be mounted "hard mounted, non-interruptible." On some hardware and networks, interrupts can cause problems when network traffic becomes heavy. When a crash such as a power failure occurs during a save, the software will detect the situation and fix the problem the next time you start - if the scratch files still exist. If you deleted them, you are out of luck. If a crash occurs during a save, immediately try to start the software again to fix the problem.

How can you protect yourself from data loss?

(1) Anybody working with computers should know that it is the responsibility of the user to back up data files, or to make sure that a system manager is in charge of doing this for you.

(2) Using libraries to store your data is also inherently safer, since part files are stored in smaller files, which are less vulnerable to corruption. If one part file becomes corrupt in the library, you have not lost the whole library, as you most likely would if all the parts were in one model file.

Software hangs or crashes

Why does the software just beep at me when I try to pick any icon?

If the software appears to be "hung" first check that it really is hung. A common user problem is that the software is asking for input in a form, but that form has inadvertently been "pushed" to the back. The software may not let you click on any other icons until you answer the question. If your cursor still moves, iconify your graphics window to see if any forms are behind it.

The software or the system is hung, how do I abort it?

First, select the "Abort" icon, which looks like a stop sign. This will abort some lengthy processes like generating an advanced shaded image.

If this doesn't work, either log in to another window or remotely from another workstation and run the script in your directory with the name STOP*. (The STOP will be followed by other characters.) You can usually run this by typing ". STOP*". This should politely stop the software and close the windows.

If crashing is a consistent problem, something is probably configured improperly. This could be the amount of swap space on your machine or the memory tuning parameters you can set from the I-DEAS preferences. To solve this problem, first see if the software has similar problems on a machine having more memory or swap space. Certain operations such as large finite element solves or hidden line processing to create drawings of complex parts or large assemblies may require special memory tuning. Be very cautious when adjusting these memory tuning values. You should not play with these virtual memory parameters without the advice of someone experienced in what these values mean.

Try to isolate the problem to see if it is limited to a particular operation, a particular machine, and/or a particular model file. If the problem is limited to a particular operation in a specific application and task, see the sections below for more specific information.

When you start to get error messages, this is NOT the time to save your model file, or any errors that might have occurred will be permanently saved in the file. When you have many errors occurring, you should re-open the file to the last save.

Software User Interface problems

This next category of problems includes problems that can occur once the software is running with menus, forms, and icons, and graphics. Problems that are specific to each application, like not being able to perform a specific construction operation or a finite element solve are discussed in later sections.

Icon/Menu/Form problems

Why doesn't an icon do anything when I click on it?

Make sure you are not holding the icon down too long. This causes the stack to "pop up" to allow you to select other icons by sliding down on the screen. If you pop up the stack, no matter how briefly, nothing will happen unless you slide the cursor down to select. The solution is either to pick more quickly, or hold longer and slide the cursor down.

Also see "Why does the software just beep at me when I try to pick any icon?" above. This can occur for example, when the software asks you if you want to save. If you click in the graphics window at that point, the message form gets hidden behind the graphics window.

There are no icons. Where did they go?

If the prompt window is iconified, the icon window goes with it. There is a preference setting that allows the icons to be turned off. Type "/option pref" if the icon panel is turned off.

Why are some icons grayed out?

Icons will be grayed out if they; don't apply for the task you are in, the hardware you are using, or for options you have set. If you are running the software with the X3D display driver, many of the hardware shading icons will be grayed out.

Why can't I turn menus on?

Menus are controlled in two places– In the *Help* menu, and in *Preferences*. In the menu preferences, you can also set three levels of menus– *Short, Long,* or *All. Short* is the default, which displays only menus that do not have an icon equivalent. If you turn menus on in the Help menu, you may not see any on the screen if all the menus in the task you are using have equivalent icons.

Why do some menus stay on the screen?

There may be places in the software where menus or submenus appear on the screen if they are options not in the icons. Menus might get turned on by a program file you run, or they might have been on all along, but only show up in certain places in the software.

Why can't I enter information in a form?

One form is active at a time. If the active form gets pushed behind an inactive form, the form on top will not accept information.

Screen graphics problems

Why can't I see anything on the screen?

There are several possible reasons why you cannot see graphics on your screen. Check *Info, Workbench* to see how many parts are on your workbench. Zoom All to bring parts off screen back on. Use Redisplay after window resizing or anytime that the graphics may need updating. Check the settings of the *Display Filter* to make sure parts or other entities are not turned off. Make sure your part has not been set to the same color as the background (such as black).

Zoom All and use F6 to reset clipping planes which may make entities invisible.

Why can't I see a particular entity type on the screen?

Check the settings in the *Display Filter*. Make sure you are in the task you should be in to manipulate the type of entity. Things could have been turned off with the *Hide* command or suppressed in the Assembly task. Use *Unsuppress* or *Show* to display them again.

Why, when I create something, does it immediately disappear?

If the entity type is turned off in *Display Filter*, it will be shown as it is created, then disappear.

Why do "old things" flicker on the screen?

When you zoom or rotate the graphics, you may see entities that have been deleted flicker on the screen. Use *Redisplay* to clear out the terminal display list.

Why can't I pick points?

Corner points on wireframe geometry vertices of parts are not pickable in hidden line mode. Use a line display to make them pickable. You may not be aware that you are still in hidden line mode if you are only working with wireframe geometry.

There is a "pickability" switch in preferences that may have been turned off.

When working with wireframe geometry, in some commands only geometry on one sketch plane is active. Geometry, even if located on the same plane in space, may be associated with a different sketch plane. Use *Attach* to move geometry from one sketch plane to another.

Why doesn't the highlighting line up with the graphics?

Select *Redisplay*. If that does not fix it, resize the window slightly, and then select *Redisplay*.

Why can't I pick or put away a part I see on the screen?

The part you see on the screen may not be a part, but an assembly instance. Switch to the Assembly task, and put away the current assembly.

Why are some lines missing on the screen?

Lines may be hidden by clipping planes, hidden with the *Hide* icon, or the *Display Selected* icon. Use F6 to reset clipping. Use *Show* to turn hidden lines back on. Use *Display All* to turn on lines temporarily turned off with the *Display Selected* icon.

Use a shaded display to make sure that your part is not missing faces.

Why can't I rotate my part on the screen?

If you started the software using X3D, you can not rotate shaded displays. Dynamic Viewing may be turned off by some operations, such as making a picture file. Execute another display to turn it back on. Dynamic Viewing can also be turned off with a preference switch.

Why do the colors "flash" to psychedelic colors?

When the color mapping table is limited on your workstation, applications will have to temporarily change the color map for the active window containing the cursor. The I-DEAS software attempts to use colors as mapped by other software. You may need to disable some desktop utilities that "rob" colors.

Why are colors missing in my display? (Especially simulation results)

Turn off double buffering in the display preferences. This option gives smoother rotation at the expense of the number of colors available.

Why does the display flicker when I rotate it?

You may have double buffering turned off. Use *Redisplay* to delete old entities in the terminal display list.

For more information on these and similar questions see

Help, Help Library
    I-DEAS User's Guide
        Global Functions in I-DEAS
            Special Functions
                Troubleshooting I-DEAS

Design Application

I deleted a part on the workbench, why is it not in the bin?

The part on the workbench is the part in the bin. It is not a copy of it. If you delete one, you delete both.

After using a part to cut another part, why is it gone?

Parts used as cutters or joined as features will be absorbed into the new part. If they have been named first, they will still be stored in the bin. If they do not have a name first, the only way to get them back is to extract the feature from the part.

Why can't I modify a part?

You can't modify a part if it is a reference from the library.

Simulation results or Optimization Designs will also prevent a part from being modified. Delete the results first if you don't want them, or make a copy of the part to preserve the results.

Make sure you are not trying to modify an assembly instance instead of a part. Put away the current assembly to make sure you know what you are working on.

Why can't I sketch in place on a part?

To use *Sketch in Place* requires the permission to modify the part, since the sketch plane will be associated with the part. If you can't modify the part, align the workplane to the face instead.

When I try to modify a part, why do my choices include dimensions but not wireframe?

You are modifying an assembly instance, not the part on the workbench. Turn off the assembly in *Display Filters*, or put it away and get the part out on the workbench.

Why does my shaded image display look "funny"?

Coplanar surfaces give a mottled effect on shaded image displays with most display hardware. Make sure you do not have two copies of the same part on the workbench.

Modified features will give the same effect until they are updated.

Why did some features disappear after I modified a part?

Features such as fillet may disappear when the part is updated if the underlying topology is deleted, or no longer exists after the update. Use *History Access* to find this features (shown as yellow or red). Fillets can be repaired using *Add Edges* and *Delete Unfound Edges*.

Why can't I fillet a part?

When you have trouble making a fillet, look for small offending geometry. Try a smaller fillet radius, or the order of how you applied the fillet with other features.

Check in the Help Library for a description of new capabilities of the fillet command.

There is a special command in the menu *Modify*, *Special Techniques* to convert old fillets to the new algorithms.

Why can't I shell a part?

There are some limitations to the geometry that the shell algorithm can handle. Avoid nearly tangent surfaces that are not constrained to be exactly tangent.

Check in the Help Library for new capabilities of the *Shell* command.

When I create an assembly constraint, why do both parts move?

Always ground one part first. This is true for other applications that use assemblies such as Manufacturing and Harness Design.

How do I know if there are assembly constraints?

Display constraint relationships between instances with the *Browse Relations* command.

How do I delete assembly constraints?

Select constraints on the *Browse Relations* form and then delete them.

For more information on these and similar questions, search for the following words.

Help, Help Library
    I-DEAS User's Guide
        Search for the words
            trouble*
            question*
            technique*

Also, find technical tips in the GTAC web site.
http://support.plms-eds.com/

Drafting Task

Why can't I create a drawing of a part on the workbench?

You must name the part before you can create a drawing.

For more information on troubleshooting with the Drafting task, see the online manual:

Help, Help Library
    Drafting User's Guide.
        Search for the words
            trouble*

Simulation Application

Why can't I create an FE model attached to a referenced part?

If the referenced part does not have a valid material for Simulation, you cannot attach an FE model to it. To modify the material requires write permission. The solution is to either:

1. Check out the part, attach a material using the *Properties* icon, and check it back in. Then reference the part.

2. Use a copy of the part. In this case, the software automatically attaches a valid default material, which can be modified or overridden in the mesh definition forms.

Why can't I mesh my part?

Make sure that the current FE model is attached to the part you have displayed on the workbench. For example, if your current FE model is attached to a null part, you cannot mesh a different part that is also on the workbench.

Other problems with mesh size settings may give specific errors when trying to mesh.

Why can't I solve my model?

Solving a model takes more computer resources of memory and disk space. Make sure you have enough disk space. Try solving a small problem first, to check for other problems. *Memory Usage* parameters may need to be set differently for your workstation.

With I-DEAS 8 or later, you may need to edit the user_param file and insert the following line if the solve does not work. This will allow you to manually change the *Memory Usage* parameters, if needed.

```
Memory.AutoSetting: 0
```

What causes a "singularity" error during the solve?

Singularity errors are caused by improper boundary conditions, improper element connections, or in some cases invalid properties.

Most solution methods require the model to be restrained to prevent all six possible rigid body motions.

If you use multiple element types, make sure they are properly connected. Since most solid elements only define three DOF, you can not cantilever a beam element at one node since beams use six DOF. This would leave the beam element free to rotate around the connection point.

Gap elements and contact surfaces do not remove singularities. If one part of a model is captivated only be gaps or contact, a singularity error will occur unless some small stiffness is included to restrain the model to allow the gap or contact can take effect.

In some cases, incorrect material or physical properties can cause singularities. For example, setting E to nearly zero would cause problems. Very small or very large thicknesses can cause the same problem. These types of problems are usually due to entering the wrong decimal location, or using the wrong units.

Why does my contour display use only three colors?

Turn off double buffering in *Preferences, Display*.

Why can't I "probe" for display results?

The *Probe* switch must be turned on in the *Display Template*.

Why are elements missing in my contour display?

Manually define the range of the color bars to include a slightly lower and higher range of values. The range of color bars is scanned based on nodal values. In some cases, higher order parabolic or cubic elements can have values between nodes can exceed the minimum or maximum values, which leaves a "hole" in the display, where the element falls outside of the plotted range.

Why can't I modify my part after solving finite element results?

Analysis results lock the model, to prevent you from displaying results on the wrong geometry and possibly making incorrect design decisions based on the displayed results.

To run another analysis, delete the old results, or make a copy of the part.

For more information, see:

Help, Help Library
Simulation Model Solution/Optimization User's Guide
Search for the word
error*

Manufacturing Application

How can I machine a referenced part if I can't rotate it into position?

Rotate the instance of the part in the setup assembly. The same holds true for adding coordinate systems. You cannot modify a reference part, but you can modify the setup assembly.

Why do I get errors creating a toolpath?

Many of the errors generated are warnings when the software changes default settings that you have not entered.

General things to check:
Is the correct Machine Coordinate System selected?
Is the correct depth entered in the Cut form?
Is the default and individual finish allowance correct?
Are there any fixtures, etc. within the avoidance allowance?
Is the tool long enough?
Is the tool nose correctly defined to fit the part?
Are correct surfaces selected?

Why doesn't the toolpath cut the stock part to show "in process" results?

In *Process Stock Calculations* must be turned on the *Setup Specification* and for each operation.

For more information, see:

Help, Help Library
Generative Machining User's Guide
Search for the word
error*

Test Application

Why can't I add or multiply two functions together?

Functions must have compatible abscissa and ordinate attributes to be processed together. They must have the same number of data points with the same spacing.

How can I edit functions based on numerical value range rather than the data point numbers?

Change the *Range Specification* switch to *Value* instead of point *Number*.

For more information, see:

Help, Help Library
Test: Basic Capabilities User's Guide

Data Management Errors

Can't read/write to a library or catalog

I checked out a part from the library, why can't I see it?

Parts or assemblies from the library are initially placed in a bin, not on the workbench. Use the Get icon to see them.

Why can't I check a part out of the library?

Only one user at a time can check a part out to modify it. If the part is currently checked out by someone else, use a reference or a copy.

Do not keep parts checked out in your model file if others need to modify the original.

How can I get a part out of a library from a different project?

You can get parts from any library in the same data installation. On the *Get From Library* form, next to the project name is a *Find* button which lets you select a different project.

Why didn't my assembly completely update when I used Update from Library?

Each subassembly level must be updated in the library. There is a new option to Update In Library to update all levels.

It is important to understand the libary status of reference specific (RFS) vs reference latest (RFL). Usually, an assembly is RFL at the top level, and RFS for lower levels.

Importing/Exporting files

Why can't I export an IGES file? (It worked before.)

Don't delete files from the operating system, or the I-DEAS data management system will continue to try to track them. Use commands in the software to delete items.

Where To Go for More Information

For more information on Data Management troubleshooting, see that section in the online Help Library.

Help, Help Library
 Data Management Guides
  Data Management User's Guide
   Troubleshooting I-DEAS Data Management (IDM)
    Troubleshooting Model File Problems
    Troubleshooting File Access Problems
    Troubleshooting Library Problems
    Troubleshooting Project Problems
    Troubleshooting Miscellaneous